"In this vitally important book, Donoso highlights how the widespread violence enacted on animals is not the result of the actions of cruel individuals, but of structural injustice. It brilliantly applies normative theorising to real-world cases to reveal the interwoven nature of the oppression of humans and animals."

Professor Alasdair Cochrane, author of *Sentientist Politics*

"Alfonso Donoso offers a powerful and elegant argument for taking the claims of other animals seriously in our moral and political life. Through an analysis of oppression, he shows that the injustices inflicted on humans and non-human animals are products of the same institutional logics. Essential reading for anyone concerned with building a more just and compassionate future."

Melanie Challenger, author of *Alive: The Hidden Intelligence of the Living World* and co-director of *Animals in the Room*

ANIMAL JUSTICE

Animal Justice argues that the systematic wrongs inflicted on billions of nonhuman animals in industrialised societies are structural injustices, embedded in the institutions, economies, and cultural norms through which human societies organise collective life.

It analyses the situation of animals not as a problem of inclusion but as one of entrenched hierarchical arrangements that obstruct the conditions necessary for self-development. Through interconnected sites – ableism and capitalist production, migration and border regimes, global commodity chains, and the everyday fabric of multispecies life – the book maps the architecture of interspecies subordination and shows how the oppression of animals is not an injustice parallel to human oppression but structurally entangled with it, produced by the same institutional logics and sustained by the same cultural norms. Against this diagnosis, the book develops an original normative response grounded in care and solidarity – understood not as personal virtues or sentimental dispositions but as demanding principles of institutional design. It asks what it would mean for political institutions to genuinely care for all those whose lives they shape, and what solidarity requires when the equal standing of animals is left to the discretion of the benevolent rather than secured as a requirement of justice. *Animal Justice* honestly confronts the possibility that institutional transformation may be blocked, arguing that when democratic arrangements persistently fail to honour the demands of justice, direct action may constitute not a rejection of political community but an act of fidelity to its ethical foundations.

Situated at the intersection of political theory, care ethics, and environmental philosophy, this book will be of interest to students and researchers in animal ethics, animal political theory, and animal studies, as well as advocates and activists for social and interspecies justice.

Alfonso Donoso is Associate Professor of Political Philosophy at the Institute of Applied Ethics and the Institute of Political Science, Pontificia Universidad Católica de Chile. His research focuses on different aspects of the political turn in animal ethics with a special interest in questions of political morality concerning animals and the political intersections between animal and environmental ethics.

ROUTLEDGE STUDIES IN ANIMALS, SOCIETY AND THE ENVIRONMENT

Vegan Witchcraft
Contemporary Magical Practice and Multispecies Social Change
Corey Lee Wrenn

Animal Justice
Confronting Oppression with an Interspecies Politics of Care
Alfonso Donoso

For more information about this series, please visit: www.routledge.com/Routledge-Studies-in-Animals-Society-and-the-Environment/book-series/RSASE

ANIMAL JUSTICE

Confronting Oppression with an Interspecies Politics of Care

Alfonso Donoso

Routledge
Taylor & Francis Group
LONDON AND NEW YORK

Designed cover image: Getty Images

First published 2027
by Routledge
4 Park Square, Milton Park, Abingdon, Oxon OX14 4RN

and by Routledge
605 Third Avenue, New York, NY 10158

Routledge is an imprint of the Taylor & Francis Group, an informa business

British Library Cataloguing in Publication Data
A catalogue record for this book is available from the British Library

ISBN: 9781032562995 (hbk)
ISBN: 9781032561653 (pbk)
ISBN: 9781003434856 (ebk)

DOI: 10.4324/9781003434856

Typeset in Sabon
by Taylor & Francis Books

CONTENTS

ACKNOWLEDGEMENTS

Through conversations and critique, many people have helped give shape to this book over the last few years. Not all of them will know they did so – some may not even know this book exists! – but their thinking, even when unrelated to the animal question, enriched this work and sharpened my own. I would like to thank Alejandra Mancilla, Pedro Riquelme, Sophie Halart, Anna Wienhues, Gaspard Donoso, Javier Correa, Dan Swain, Harald Borgebund, John Charney, Juan Larraín, Rosine Kelz, Jessica Gimpel, Tomás Chuaqui, Sasha Mudd, Constanza Guajardo, Ross Mittiga, Frederike Spang, Carolina Leiva, Rodrigo Mardones, Julita Skotarska, Matt Matravers, Gabriela Arriagada, Christof Mauch, Angie Pepper, Juan Pablo Luna, Iwona Janicka, Alasdair Cochrane, Richard Healey, Petr Urban, Avery Kolers, Josh Milburn, Juan Pablo Mañalich, Cristian Pérez, Markku Oksanen, Olof Page, and Sevgi Mutlu Sirakova.

I would also like to remember Siobhan O'Sullivan, whose work on animal politics has been important to this project. We met only once, at a conference in Switzerland, but the conversations we had on that occasion and the written exchanges that followed were profoundly inspiring. The field is diminished by her absence.

My academic homes, the Institute of Political Science and the Institute of Applied Ethics at the Pontificia Universidad Católica de Chile, have provided a generous and sustaining environment in which to pursue this work. Earlier versions of chapters and arguments contained in this book were presented at the Rachel Carson Center for Environment and Society in Munich, the Center for Environmental and Technology Ethics in Prague, the Helsinki Animal Law Centre, the Department of Philosophy, Classics, History of Art and Ideas at the University of Oslo, Østfold University College, and the University of Chile. I am grateful to the participants on all these occasions for their comments and engagement.

I am also grateful to Serena Watkins and Hannah Studd at Routledge for their support, patience, and guidance throughout the process, and to three anonymous reviewers whose comments and objections strengthened the project considerably. Parts of this work were supported by ANID (Fondecyt N° 11160170 and Fondecyt N° 1260474) and by a fellowship from the Rachel Carson Center for Environment and Society at Ludwig-Maximilians-Universität München, which allowed me to spend several months working in Munich during my sabbatical alongside its remarkable community of scholars.

Above all, I would like to thank Sophie, Islay, Gaspard, Élise, and Patou. Each of them, in their own ways, at their own times, has marked this book and much more beyond it.

1

JUSTICE BEYOND HUMANS

1 Sources of Resistance

When I first began thinking about the relationship between humans and other animals, what struck me was not the difficulty of the moral questions themselves, but the resistance to following them wherever they lead. There is, across academic disciplines and in broader public culture, a widespread reluctance to engage not merely with whether animals have moral status – that question, for many, is already settled – but with what that status actually requires of us. The harder and more consequential question is what follows from taking animal value seriously, and it is precisely here that inquiry tends to stall.

This resistance is not simply a matter of indifference or cruelty. It takes several distinct forms, some more philosophically interesting than others. One familiar objection holds that taking the moral claims of animals seriously must come at the expense of what we owe to fellow human beings – as though moral attention were a finite resource, and extending it to animals necessarily depletes what remains for humans. This is a mistake, and an instructive one. There is no principled reason why recognising the value of one thing diminishes our reasons to value another. If anything, the habits of moral attention that serious engagement with animal ethics demands – attentiveness to suffering, resistance to self-serving rationalisations, willingness to revise entrenched practices – are precisely the habits that strengthen rather than erode our moral relationships with one another.

A second source of resistance concerns urgency. In a world convulsed by war, genocide, forced displacement, and the accelerating erosion of democratic institutions, what could justify devoting serious intellectual attention to the situation of animals? The objection has genuine force. It would be a form of

DOI: 10.4324/9781003434856-1

moral obtuseness to be indifferent to the scale and immediacy of human suffering, and no serious animal ethics can afford to proceed as though that suffering does not exist or does not press upon us with particular urgency.

And yet the objection rests on a questionable premise – that the urgency of some forms of suffering renders others unworthy of sustained inquiry. This is, again, the zero-sum assumption in a different guise. The crises that rightly command our attention do not suspend the claims of those who suffer beyond their immediate frame. Moreover, the suffering inflicted on animals is not marginal in scale: it is among the largest and most institutionally entrenched forms of harm, affecting trillions of sentient lives across every corner of the globe. If urgency is the criterion, the animal question does not obviously fail it.

There is a further point. The political and economic structures that organise the exploitation of animals are not separate from the structures that produce the human crises the objection invokes. They share logics – of disposability, of the conversion of living beings into resources, of the subordination of the vulnerable to the interests of the powerful. A political theory capable of illuminating one may, for that very reason, have something important to say about the other. This book argues that the connection runs deeper than analogy. The chapters that follow show that the oppression of animals and the oppression of human beings are not parallel injustices – they are structurally entangled, produced by the same institutional logics and sustained by the same cultural norms. To take the animal question seriously is not to turn away from human oppression. It is to see that oppression more completely.

A third source of resistance is more philosophically specific and in some ways more interesting than the first two. It holds that while animals may be harmed, exploited, or killed, they cannot properly be described as being treated unjustly. Justice, in this view, is a concept with determinate conditions of application: it presupposes the kind of agency, reciprocity, and normative standing that make beings capable of having claims, not merely interests, against others and against social institutions. Animals, lacking these features, may be objects of moral concern but not subjects of justice.

This objection deserves to be taken seriously, and this book takes it seriously. It is not merely a defensive manoeuvre by those indifferent to animal suffering. Some of its most careful versions come from thinkers genuinely committed to animal protection but who believe that the political vocabulary of justice does not translate across the species boundary without distortion. The question they raise is real: what work is the concept of injustice doing when applied to a broiler chicken or a laboratory macaque, and does it survive that application intact?

This book answers that it does, but only if we attend carefully to what kind of injustice is at stake. The book argues that animals are not merely harmed but oppressed, and that oppression names a specific and particularly serious form of injustice: one grounded not in individual wrongdoing but in structural arrangements that systematically deny entire groups the conditions necessary for self-development. The most influential modern account of this form of injustice, developed by Iris Marion Young (1990, 2011), does not ground oppression in rational agency or political consciousness. It grounds it in something more basic: the systematic obstruction of the conditions under which beings can develop and exercise their capacities, sustained by institutions, cultural norms, and economic systems that sort beings into hierarchies of value. What makes a wrong oppressive, in this account, is not the victim's capacity for political reflection but the structural character of the harm, its embeddedness in social practices that reproduce it independently of individual malice or intent. This is a conception capacious enough to reach beyond the human, and the chapters that follow show in detail how and why it does so.

The fourth source of resistance follows naturally from the third. If animals are subjects of injustice, and if that injustice is structural rather than merely a matter of individual cruelty or moral failure, then the appropriate response cannot be confined to welfare reform or the restraint of individual behaviour. Structural injustice requires structural responses. That means directing attention toward the institutions – legal, economic, and political – that produce and sustain animal oppression, and asking what it would mean to transform them. This is where the deepest resistance tends to emerge, because the implications are genuinely radical. It is not simply a matter of regulating or reforming slaughterhouses more humanely or extending modest legal protections to certain species. It is a question of how political authority is organised and exercised, of who counts as a subject of governance, and of whether existing arrangements – including the property regimes, the food systems, the urban infrastructures, and the democratic institutions through which we govern collective life – can be made compatible with animal justice or must instead be fundamentally reconceived. Many who are willing to accept that animals suffer, and even that their suffering matters morally, find this implication too demanding. This book argues that the demand is nonetheless inescapable once the structural character of animal oppression is properly understood.

2 Animal Ethics

Systematic philosophical engagement with the moral status of animals is, by historical standards, surprisingly recent. For most of the Western philosophical tradition, the question was either dismissed outright or confined to the margins of ethical inquiry. Descartes treated animals as automata, incapable of genuine experience; Kant excluded them from the moral community because they lacked rational agency; and even utilitarian thinkers who acknowledged animal suffering rarely followed that acknowledgement into sustained moral or political theorising. The emergence of animal ethics as a recognised philosophical subdiscipline is largely a product of the last half-century, and it begins, in its modern form, with two works that continue to define the terms of debate.

Peter Singer's *Animal Liberation* (1975) made the first decisive intervention. Drawing on utilitarian foundations, Singer argued that the capacity for suffering – not intelligence, rationality, or species membership – is the morally relevant criterion for inclusion within the scope of ethical concern. To discount the suffering of animals simply because they are not human is, on his account, a form of prejudice analogous to racism and sexism, a bias called speciesism. Singer's argument was not primarily theoretical; it was a call to action, grounded in detailed documentation of the conditions of factory farming and animal experimentation. Its impact on public consciousness and on the subsequent development of the field was profound.

Tom Regan's *The Case for Animal Rights* (1983) offered a philosophically distinct but complementary challenge. Where Singer grounded moral consideration in sentience, Regan argued from a rights-based framework: animals who are, in his phrase, subjects-of-a-life – beings with beliefs, desires, perception, memory, and a welfare that matters to them independently of its utility to others – possess inherent value that cannot be overridden by appeals to aggregate benefit. For Regan, the problem with how we treat animals is not merely that we cause them unnecessary suffering but that we treat them as means rather than ends, violating rights that are theirs by virtue of what they are.

Together, Singer and Regan established the two major poles around which animal ethics has since organised itself: a utilitarian tradition focused on the reduction of suffering, and a rights-based tradition focused on the inviolability of individual animals. Despite their differences, both shared certain assumptions that would later come under pressure. Both worked primarily within moral philosophy rather than political theory. Both focused on individual animals and their experiences rather than on the structural conditions that produce and sustain harm at scale.

3 Political Theory and Beyond

The most significant recent development in animal ethics has been what is now usually described as the political turn (e.g., Milligan, 2015). Dissatisfied with the limitations of purely moral approaches, a growing number of theorists have argued that the situation of animals cannot be adequately addressed through ethics alone. What is needed, they contend, is a genuinely political theory of human–animal relations, one that analyses animals not merely as objects of moral concern but as members of political communities whose relationships with humans generate specific obligations grounded in justice.

The landmark text of this turn is Sue Donaldson and Will Kymlicka's *Zoopolis* (2011). Drawing on political philosophy rather than moral theory, Donaldson and Kymlicka propose a tripartite framework organised around the different relationships that exist between humans and different categories of animals. Domesticated animals, they argue, should be understood as co-citizens of the political communities in which they live, entitled to the rights and protections that citizenship implies. Wild animals are sovereign communities whose territories and self-governance deserve respect analogous to that owed to independent nations. And liminal animals – those who live alongside humans without being domesticated – occupy a position analogous to resident aliens, with corresponding claims to accommodation and non-interference. *Zoopolis* was significant not only for the substance of its proposals but for its methodological move: treating animals as political subjects rather than moral patients, and asking what political theory, rather than ethics, requires of us.

Other contributions to the political turn have developed adjacent but distinct frameworks, collectively demonstrating that the politicisation of animal ethics is no longer a marginal provocation but an established and diversifying research programme. Alasdair Cochrane has examined the political rights of animals from a liberal perspective, arguing for a robust set of entitlements grounded in sentience and interests rather than agency or rationality (Cochrane, 2010, 2012, 2018). His work is distinctive in its insistence that interests alone are sufficient to generate enforceable political claims, without requiring controversial assumptions about animal cognition or participation. Cochrane also highlights the limitations of welfare reforms and legal personhood, advocating for democratic representation as an adequate mechanism for ensuring animals' interests are meaningfully included in political decision-making – a species-neutral, cosmopolitan vision in which animals figure as political subjects entitled to protection and inclusion. Robert Garner has explored the institutional and democratic dimensions of animal politics, asking what kind of state would be required to take animal interests seriously and how existing democratic arrangements might be reformed or extended to include nonhuman concerns (Garner, 2013). Clare Palmer has advanced a relational approach that grounds obligations to animals in the particular histories and dependencies that characterise different human–animal relationships (Palmer, 2010). Siobhan O'Sullivan (2011) brought a comparativist and policy-oriented perspective to animal politics, examining how different regulatory regimes across liberal democracies produce systematically inconsistent treatment of animals, protecting high-visibility animals more effectively than low-visibility ones. She argues that equitable lawmaking – applying consistent standards regardless of public exposure to animal suffering – is both achievable within liberal democratic frameworks and a necessary condition of legitimate animal governance.

More recent contributions have extended the political turn in important new directions. Steve Cooke (2017) has argued that achieving justice for animals requires cultivating the moral imagination – the capacity to envision radically different arrangements – and that this cultivation depends on institutional conditions: public support for the arts and humanities, freedom of thought, and philosophical work that expands the conceptual space within which alternative futures can be imagined. Eva Meijer (2019), pursuing a distinct line of inquiry, draws on language and political theory to argue that animals have political voice, and that a genuinely democratic politics must develop institutional mechanisms capable of incorporating these voices into processes of collective decision-making. Angie Pepper (2021) has developed a rigorous account of political agency that identifies the capacities required for genuinely political action, concluding that most nonhuman animals, human infants, and humans with severe cognitive impairments cannot straightforwardly be assigned rights to political participation – a finding that, on the present account, sharpens rather than forecloses the question of what political standing animals can nonetheless claim. Josh Milburn (2023) has examined what a rights-respecting food system would look like, challenging the assumption that animal justice requires a fully plant-based diet and arguing that there are principled reasons to prefer a non-vegan but rights-respecting alternative that includes cultivated meat, non-sentient invertebrates, and carefully managed animal husbandry. Pablo Magaña (2022) has made important contributions to the question of political representation for animals, systematically examining the different institutional mechanisms – from ombudspersons to designated legislative representatives – through which the interests of nonhuman animals might be incorporated into democratic governance without either anthropomorphising their agency or reducing their claims to mere welfare considerations. Together, these and other contributions show how far the political turn in animal ethics has developed beyond its founding statements into a differentiated and institutionally ambitious research programme.

A substantial body of work at the intersection of critical theory, feminist philosophy, and postcolonial thought analyses animal exploitation through the lenses of power, race, gender, and capitalism, connecting the subjugation of animals to broader structures in ways that purely liberal frameworks tend to obscure. Carol J. Adams's *The Sexual Politics of Meat* (1990) demonstrated the structural links between patriarchy and the exploitation of animal bodies. Aph and Syl Ko's *Aphro-ism* (2017) argued that the concepts of "the human" and "the animal" are constructed through racialised taxonomies of power, that anti-Blackness and speciesism share a common logic of dehumanisation, and that genuinely liberatory politics must dismantle both simultaneously – developing black veganism and Afrofuturist frameworks as alternatives to Eurocentric understandings of race and species. Dinesh Wadiwel's *The War Against Animals* (2015) has drawn on Foucault, Derrida, and Marx to argue that human–animal relations constitute a condition of permanent war in which sovereign claims of human superiority are founded on violence alone, that property in animals represents a biopolitical conquest rather than a neutral legal arrangement, and that pro-animal politics must challenge this violent sovereignty by recognising animal resistance and working toward forms of counter-conduct and truce.

Yet even this critical tradition has remained, for the most part, within the conceptual orbit of Western thought. This is a significant limitation. Long before animal ethics emerged as a philosophical subdiscipline, many Indigenous and non-Western traditions had developed sophisticated and politically consequential frameworks for understanding the relationships between humans and other animals – frameworks that do not begin from the assumption of human exceptionalism that underlies so much of the Western canon.

In many Indigenous cosmologies across the Americas, Australasia, and Africa, animals are understood not as objects of moral concern but as persons, kin, or fellow members of relational communities whose claims on humans are grounded in shared history, reciprocity, and obligation rather than in abstract rights or utilitarian calculation. Robin Wall Kimmerer's account of Potawatomi thought, developed in *Braiding Sweetgrass* (2013), articulates a conception of reciprocity between humans and the living world structured around gratitude, obligation, and mutual gift rather than ownership and use – centred on plants but extending to all animate beings – and grounded in the recognition that the natural world is composed of persons, not resources. In many First Nations traditions of North America, hunting is governed not by the logic of resource extraction but by protocols of respect, consent, and reciprocal obligation that presuppose the animal's personhood and the hunter's accountability to the community of life. Similar structures of multispecies relationality appear in Andean cosmologies of *sumak kawsay* (*Buen Vivir*), in which the flourishing of the community is inseparable from the flourishing of the land and its nonhuman inhabitants (e.g., Acosta, 2012; Altmann, 2017), and in many African Ubuntu-inflected frameworks that extend the logic of relational personhood beyond the human (e.g., Diagne, 2023; Ewuoso & Hall, 2019).

These traditions are not merely philosophical curiosities or ethnographic data points. They represent genuinely different ways of organising the political relationships between humans and other animals, and they carry practical implications that Western animal ethics has only begun to register. They also offer an important corrective to the implicit universalism of the Western tradition – the tendency to treat as a novel philosophical discovery what many other traditions have long understood and institutionalised. In many parts of the world, that question has been asked – and partially answered – for centuries, through practices and institutions that Western modernity has systematically marginalised or destroyed.

This book does not claim to offer a synthesis of these traditions. Its primary theoretical resources remain within the political philosophy of the Western academy, and it makes no pretence otherwise. But it proceeds in the awareness that this is a limitation rather than a neutral starting point, and that a fully adequate theory of interspecies justice will eventually need to engage seriously with the forms of multispecies political thought that Western philosophy has too long ignored.

Taken together, these contributions have transformed what began as a provocative extension of political philosophy into a rich and contested field, one in which questions of justice, democracy, sovereignty, and institutional design are now being asked seriously and in detail about beings that mainstream political theory has historically ignored entirely. This book builds on that field and is indebted to it. If it makes a contribution to the political turn in animal ethics, it does so by extending foundations that others have laid, and in the hope of continuing a conversation rather than closing one.

4 Where This Book Stands

This book belongs to the political turn, but it occupies a specific and, in some respects, critical position within it. It shares with the political turn broadly the conviction that the situation of animals is a matter of justice rather than mere benevolence, and that political theory rather than moral philosophy alone is the appropriate framework for analysing it. It shares with the participatory and democratic strand of that tradition the view that animals are not merely passive recipients of human decisions but beings whose lives, capacities, and relationships generate genuine political claims. And it shares with the critical strand – attentive to race, gender, colonialism, and structural power – the recognition that animal exploitation cannot be understood in isolation from the broader structures of hierarchy and subordination that organise social life.

But the book departs from much of the existing literature in two important and connected respects.

The first concerns the theoretical framework used to analyse animal injustice. Much of the political turn, including its most influential contributions, has worked within broadly liberal frameworks – extending rights, citizenship, and sovereignty to animals, and asking what existing political concepts require when applied beyond the human. This is valuable and necessary work. But it tends to approach the situation of animals as a problem of inclusion: of bringing previously excluded beings within the scope of principles already established for humans. What it is less well equipped to do is analyse the structural conditions that produce and reproduce animal subjugation in the first place – the institutional arrangements, economic logics, cultural norms, and political hierarchies that organise the exploitation of animals not as a series of discrete wrongs but as a systemic and self-perpetuating order.

This book addresses that gap by drawing on Iris Marion Young's account of structural oppression. Young's framework allows us to see animal injustice not as a failure of inclusion but as a structural problem – one that is embedded in our institutions, economies, and ways of life, and that operates independently of individual malice or intent. The central claim is that animals are not merely harmed but oppressed, and that oppression names a specific and particularly serious form of injustice: the systematic obstruction, through entrenched group hierarchies and structural arrangements, of the conditions necessary for self-development. Applying Young's five faces of oppression – exploitation, marginalisation, powerlessness, cultural imperialism, and violence – to the situation of animals in industrialised societies reveals a coherent architecture of interspecies subordination that neither moral philosophy nor liberal political theory has fully captured.

The need for a more precise normative vocabulary can be illustrated through Martha Nussbaum's recent contribution to animal justice. In *Justice for Animals* (2023), Nussbaum draws a fundamental distinction between harms, wrongs, and injustice: not every harm constitutes an injustice; only some harms are wrongs; and only some wrongs are unjust. Understanding what makes a wrong an injustice requires attending to the conditions under which meaningful striving is obstructed. Injustice, on Nussbaum's account, involves the negligent or deliberate thwarting of significant striving – not mere misfortune or accidental harm, but the blocking of a life's forward movement by something or someone that could and should have acted otherwise (Nussbaum, 2023, pp. 4-8). This is a genuine and important conceptual advance, and this book builds on it.

But it does not go far enough. Nussbaum's framework is well equipped to identify individual wrongs and to ask who bears responsibility for remedying them. What it is less equipped to do is capture the specifically structural character of the wrongs animals suffer – the fact that their thwarted striving is not the result of isolated negligent or deliberate acts but of entrenched institutional arrangements that operate independently of individual intent, reproduce themselves across time, and organise entire categories of beings into permanent positions of subordination. A broiler chicken's inability to move, a laboratory macaque's confinement, a jaguar displaced by habitat fragmentation – these are not simply cases of meaningful striving blocked by identifiable wrongdoers. They are the predictable outputs of systems institutionally organised and culturally normalised to produce exactly these outcomes. To name that kind of wrong adequately, we need the concept of oppression – and it is to that concept that this book turns.

The second departure concerns the normative response that structural analysis demands. Diagnosing oppression is only the first step; what follows must be a theory of how to respond to it, and that theory must match the structural character of the problem. Rights-based frameworks, however important, tend to operate at the level of individual entitlements and their protection, and while this is necessary, it is insufficient for addressing the structural arrangements that generate those entitlements' systematic violation in the first place. What is needed, in addition, is an account of the institutional principles and political values capable of guiding the transformation of the arrangements that produce oppression in the first place. Here the book draws on the ethics and politics of care, arguing that care and solidarity – understood not as personal virtues or sentimental dispositions but as demanding institutional principles – provide the normative foundation for a genuinely transformed interspecies politics. The connection between the two departures is not incidental: it is precisely because oppression is structural, reproduced through institutional arrangements rather than individual acts, that the response must be institutional in kind – and it is care and solidarity, conceived as principles of institutional design rather than personal conduct, that supply the normative content that structural critique alone cannot provide.

This combination of structural critique and care-based normativity gives the book its distinctive shape. The first part develops the diagnosis across four sites of structural injustice; the second elaborates the normative response through institutional principles of care and solidarity; and the third confronts the possibility that institutional transformation may be blocked, asking what remains legitimate when democratic arrangements persistently fail to honour the demands of justice.

This is, in the end, a book about the kind of political community we inhabit and the kind we ought to build. Its argument is that we cannot build a just political community – for humans or for animals – without confronting the structural arrangements that currently make justice impossible for both. That confrontation is uncomfortable. It requires not merely recognising moral worth but transforming the institutions through which power is organised and exercised, and asking honestly what remains permissible when those institutions refuse to be transformed. This book is an attempt to show why that confrontation is necessary, what it would involve, and why, despite everything, it remains possible.

5 Overview

The book is organised in three parts that follow a deliberate argumentative progression: from structural diagnosis, through normative response, to the question of what remains possible when that response is institutionally foreclosed.

Part I, *Spheres of Oppression*, develops the central diagnostic claim across four distinct but interconnected sites of structural injustice. Chapter 2 lays the theoretical foundation by arguing that nonhuman animals can be subjects of oppression in a fully political sense. Drawing on Iris Marion Young's account of structural injustice, it identifies the defining features of oppression – its grounding in entrenched group hierarchies and its systemic character, sustained by institutions, laws, and cultural norms rather than by discrete acts of wrongdoing – and shows, through a systematic application of Young's five faces of oppression, how animals in industrialised societies are structurally positioned as subordinate and disposable.

Chapter 3 examines the entanglement of ableism, capitalism, and speciesism, arguing that ableism is not a marginal prejudice but a core logic through which capitalist societies organise value, productivity, and disposability across species lines. Focusing on contemporary poultry production, it shows how analogous evaluative standards govern the treatment of both humans with disabilities and nonhuman animals. It introduces the concept of anticipatory ableism to capture a distinctively interspecies dynamic. Rather than merely encountering pre-existing traits through unaccommodating structures, production systems actively engineer bodies toward particular capacities and then classify those that fall short as deficient. The chapter further shows how technological innovations in this domain frequently modernise and intensify oppression rather than overcoming it.

Chapter 4 turns to migration and borders, analysing how the political and legal infrastructures that endanger human migrants simultaneously restructure ecosystems and generate parallel forms of vulnerability for nonhuman animals. Migrant workers confined to slaughterhouses, migratory birds disrupted by hardened border regimes, and jaguars fragmented from their habitats by border infrastructure – or drawn into illegal pet trade networks – are shown to be implicated in the same structures of disposability. The US–Mexico border wall serves as a paradigmatic site in which these convergences are rendered visible: a political artefact that functions simultaneously as a locus of human rights abuse and ecological destruction, binding human and nonhuman lives within a shared regime of structurally produced vulnerability.

Chapter 5 examines global justice, showing how colonial legacies, labour exploitation, cultural imperialism, and ecological destruction converge within transnational commodity chains. An analysis of the Malaysian palm oil industry brings together the forced sedentarisation of the Batek, the persistence of child labour, and the systematic displacement and killing of Bornean elephants as interconnected expressions of a global economic order organised around extractivism and export-led growth – and responsibility for these harms is traced, via Young's social connection model, to all those whose ordinary economic practices help reproduce the background conditions under which such oppression persists.

Part II, *Toward Interspecies Justice*, develops the book's normative response to the diagnosis established in Part I. If oppression is structural, the response must be institutional, which requires articulating the values and principles capable of guiding institutional transformation. Chapter 6 argues that care and solidarity, understood not as personal virtues but as demanding political principles, provide the normative foundation for a genuinely transformed interspecies politics. It shows how institutions can embody care in a meaningful, non-interpersonal sense – by structuring recognition, responsibility, and provision for all inhabitants of a multispecies polity – and how solidarity requires accepting shared vulnerability and collective responsibility across species lines.

Chapter 7 examines the constitutional dimensions of interspecies justice, asking what kind of fundamental law a multispecies polity would require. It argues that constitutional silence on the moral standing of animals amounts to an implicit endorsement of anthropocentric exclusion, and examines how existing constitutional frameworks in countries including Ecuador, Germany, and India have begun – partially and imperfectly – to contest that exclusion.

Chapter 8 applies the principles developed in the preceding chapters to food systems, arguing that arrangements that require oppression to function cannot be rendered just through compensatory reform, and that the emergence of alternative food systems – from sanctuary farming and agroecology to plant-based and cultured meat – shows that animal exploitation is not a structural necessity but a political choice.

Chapter 9 turns to the city as a site of interspecies justice, examining how urban design, environmental policy, and civic culture either reproduce or contest the conditions of animal oppression, and articulating what a genuinely caring and solidaristic city would require in terms of spatial organisation, metabolic flows, and civic dispositions oriented toward multispecies coexistence.

Part III, *Justice Beyond Institutions*, directly confronts the possibility that institutional transformation may be blocked – where democratic arrangements, instead of enabling interspecies justice, serve to normalise, legitimise, and entrench animal oppression. Chapter 10 explores direct action as a form of interspecies political responsibility under conditions of democratic foreclosure. It argues that when institutional mechanisms no longer offer a viable path to justice, when oppression is maintained through diffuse structures that evade accountability, and when legal frameworks protect injustice rather than prevent it, citizens may bear a political responsibility to act outside or even against prevailing norms. The chapter outlines three conditions under which such action may be legitimate: democratic foreclosure, structural complicity, and restorative orientation. It also addresses the contested issue of violence, suggesting that force directed at the material infrastructures of oppression can, under specific and demanding conditions, align with a forward-looking commitment to repair and multispecies solidarity.

Chapter 11 synthesises the book's central arguments through three interlocking stories – of a Guatemalan migrant worker, a group of laboratory macaques, and a military training pig – showing that the oppression of humans and animals is not a set of parallel injustices but a structurally entangled one, produced by the same institutional logics and sustained by the same cultural norms. It also acknowledges two unresolved tensions the book cannot fully settle: how to adjudicate genuine conflicts between human and animal interests, and how to avoid paternalism in any human-led project of interspecies justice.

The chapters that follow do not claim to resolve the questions they raise. Political philosophy rarely settles what it opens up, and this book is no exception. What it can do – what it attempts to do – is take those questions seriously enough to follow them wherever they lead, even when the destination is uncomfortable and the path is contested.

I am writing at a moment of compounding crises. Ecological systems are collapsing at a pace that outstrips the capacity of existing institutions to respond. Democratic arrangements in many parts of the world are eroding under the pressure of authoritarian politics, corporate capture, and the systematic exclusion of vulnerable populations from meaningful political participation. And the scale of animal exploitation continues to expand – more animals confined, more habitats destroyed, more species pushed toward extinction – while the political and legal frameworks that might contest these developments remain largely organised around the assumption that nonhuman animals are resources to be managed rather than subjects to be protected.

It is tempting, in this context, to treat the question of animal justice as a luxury – a concern for stable and prosperous societies that have already resolved more pressing human problems. This book tries to show why that temptation should be resisted. The oppression of animals is not separate from the crises that surround us. It is entangled with them – sharing their structural logics, feeding their institutional failures, and reproducing the same hierarchies of disposability that organise the treatment of the most vulnerable human communities. To take animal justice seriously is not to turn away from the world's urgent problems. It is to see those problems more clearly, and to understand that the political imagination capable of addressing them must be expansive enough to include all sentient beings whose lives are shaped by the arrangements we have built and continue to sustain.

This book is an attempt to contribute to that expanded political imagination. It does not offer a blueprint or a programme. What it offers instead is a framework for seeing animal oppression as structural, for understanding what justice would require in response, and for thinking honestly about what remains possible when the institutions tasked with delivering justice fail to do so. If it persuades its readers that the question of how we organise our relationships with other animals is not a marginal philosophical curiosity but a central problem of political life, it will have achieved its purpose. The rest, as with all questions of justice, depends on what we are collectively willing to do.

References

Acosta, A. (2012). *Buen vivir. Sumak kawsay: Una oportunidad para imaginar otros mundos*. Abya Yala.

Adams, C. J. (1990). *The sexual politics of meat: A feminist-vegetarian critical theory*. Continuum.

Altmann, P. (2017). Sumak Kawsay as an element of local decolonization in Ecuador. *Latin American Research Review*, 52(4), 749–759. doi:10.25222/larr.242.

Cochrane, A. (2010). *An introduction to animals and political theory*. Palgrave Macmillan.

Cochrane, A. (2012). *Animal rights without liberation: Applied ethics and human obligations*. Columbia University Press.

Cochrane, A. (2018). *Sentientist politics: A theory of global inter-species justice*. Oxford University Press.

Cooke, S. (2017). Imagined utopias: Animal rights and the moral imagination. *Journal of Political Philosophy*, 25(4), e1–e18. doi:10.1111/jopp.12136.

Diagne, S. B. (2023). *Ubuntu: Entretien avec Françoise Blum*. Éditions de l'EHESS.

Donaldson, S., & Kymlicka, W. (2011). *Zoopolis: A political theory of animal rights*. Oxford University Press.

Ewuoso, C., & Hall, S. (2019). Core aspects of Ubuntu: A systematic review. *South African Journal of Bioethics and Law*, 12(2), 93–103. doi:10.7196/SAJBL.2019.v12i2.679.

Garner, R. (2013). *A theory of justice for animals: Animal rights in a nonideal world.* Oxford University Press.

Kimmerer, R. W. (2013). *Braiding sweetgrass: Indigenous wisdom, scientific knowledge and the teachings of plants.* Milkweed Editions.

Ko, A., & Ko, S. (2017). *Aphro-ism: Essays on pop culture, feminism, and Black veganism from two sisters.* Lantern Books.

Magaña, P. (2022). The political representation of nonhuman animals. *Social Theory and Practice*, 48(4), 665–690.

Meijer, E. (2019). *When animals speak: Toward an interspecies democracy.* New York University Press.

Milburn, J. (2023). *Food, justice, and animals: Feeding the world respectfully.* Oxford University Press.

Milligan, T. (2015). "The Political Turn in Animal Rights," *Politics and Animals* 1 (1): 6–15.

Nussbaum, M. C. (2023). *Justice for animals: Our collective responsibility.* Simon & Schuster.

O'Sullivan, S. (2011). *Animals, equality and democracy.* Palgrave Macmillan.

Palmer, C. (2010). *Animal ethics in context.* Columbia University Press.

Pepper, A. (2021). Political agency in humans and other animals. *Contemporary Political Theory*, 20 (2), 296–317. doi:10.1057/s41296-020-00430-4.

Regan, T. (1983). *The case for animal rights.* University of California Press.

Singer, P. (1975). *Animal liberation.* New York Review/Random House.

Wadiwel, D. (2015). *The war against animals.* Brill.

Young, I. M. (1990). *Justice and the politics of difference.* Princeton University Press.

Young, I. M. (2011). *Responsibility for justice.* Oxford University Press.

PART I

Spheres of Oppression

2

THE OPPRESSION OF ANIMALS

Structural Injustice Beyond Humans

1 Introduction

It is past midnight, and Carolina continues to pack cereal bags into cardboard boxes at a relentless pace. She must work quickly to meet the demands of the company that employs her, and carefully, so as not to injure her hands. She is just 15 years old. A year earlier, in 2021, she undertook the arduous journey alone from Guatemala to the United States, seeking employment to support her ailing grandmother (Dreier, 2023). Carolina's story is neither anomalous nor exceptional; it is emblematic of a broader and deeply entrenched pattern of child labour and forced migration that traverses the Americas and extends across the globe. At this very moment, thousands of children undertake similar journeys, paths marked by precarity, exploitation, and neglect.

This story reveals a darker side of human relations. The physical and emotional trauma Carolina has endured is severe and may leave lasting scars on her and her family. Yet the wrong she suffers cannot be fully understood by attending only to the cruelty of her supervisors or the violence of the smugglers who transported her across borders. To grasp the full moral weight of her situation, we must look beyond individual wrongdoing to the broader social and political context that enables it. The conditions under which Carolina labours are not accidental. They are produced and sustained by an economic order that systematically generates vulnerability and dependence, while concealing the hierarchies of privilege from which others benefit. Carolina is not merely harmed; she is oppressed.

In another part of the world, at the Wat Tham Pha Mak Ho monastery in northern Thailand, the population of rhesus macaques has plummeted by the hundreds. The cause remains uncertain, but all evidence points to animal trafficking. Since the onset of COVID-19, the smuggling of long-tailed

DOI: 10.4324/9781003434856-3

macaques has surged dramatically. Each monkey has a unique story, yet their fate is tragically uniform. After roaming freely around the monastery, they are captured by traffickers, stuffed into bags, and transported to holding farms. Some die quickly – life in confinement is miserable. Others are sent to breeding facilities in Cambodia, where they receive falsified papers before being shipped to the United States, destined for biomedical laboratories. There, they will live and die in sterile cages, used for pharmaceutical testing and discarded when no longer useful (US Department of Justice, 2022).

As with Carolina, the wrongness of the macaques' condition cannot be fully explained by pointing to the cruelty of smugglers or the conduct of laboratory technicians. To stop there would be to mistake symptoms for causes. These macaques are enmeshed in global systems of extraction, trade, and experimentation that mirror the very structures enabling Carolina's exploitation. Their suffering, like hers, is the predictable outcome of an institutional and cultural order that privileges certain forms of life while reducing others to instruments. The macaques, too, are victims of oppression.

To describe animals as oppressed is an unconventional, and for many unsettling, claim. The term oppression has long belonged to the lexicon of human politics, used to name the injustices faced by workers, women, racialised groups, and colonised peoples. When extended to animals, it is often met with scepticism. Critics argue that animals may be harmed, abused, or killed, but not oppressed. The concept, they contend, presupposes features animals lack: rational awareness, political agency, or group identity.

These objections are not confined to explicit speciesism. Even among those who defend stronger moral protections for animals, there remains hesitation to adopt the language of politics and justice. Yet, as this chapter argues, this hesitation forecloses more than it reveals. To understand the full scope of the wrongs inflicted on animals in contemporary societies, we must move beyond moral concern for their suffering – however genuine – for it leaves the structural conditions that produce that suffering entirely unexamined, and recognise instead the political dimensions of their subordination.

What follows develops this by drawing on Iris Marion Young's influential account of structural oppression (Young, 1990, 2000). Young's framework allows us to distinguish between individual acts of cruelty and the systemic patterns that organise social hierarchies. While her work has been discussed primarily within human-centred political theory, this chapter shows that its conceptual architecture – its focus on structure, group-based hierarchy, and the obstruction of self-development – can be fruitfully extended to nonhuman animals. Building on the insights of Lori Gruen (2009) and Azizjon Bagadirov (2025), the chapter critically extends these efforts, highlighting the need for a fully political, nonspeciesist understanding of animal subjugation.

The chapter proceeds in four steps. The next section revisits Young's analysis of oppression, clarifying its core features and the wrong it entails, develops the account of self-development through Bagadirov's concept of developmental enclosure, and establishes sentience as the threshold condition for the framework's application. The third section addresses the question of whether animals can meaningfully be said to be subjects of oppression, considering the most common objections. The fourth section applies Young's five faces of oppression – exploitation, marginalisation, powerlessness, cultural imperialism, and violence – to animals, showing how these categories illuminate the structural injustices that define their lives in industrialised societies. The chapter concludes by situating this analysis within the broader argument of the book.

2 From Cruelty to Structural Injustice

To understand what connects Carolina's story and that of the macaques, we need a conceptual language adequate to the scale and depth of the injustices they endure. Neither case can be plausibly explained in terms of individual cruelty, moral failure, or misfortune. The harms at issue are sustained by vast, interlocking economic, political, and cultural systems that sort beings into hierarchies of value – systems that determine whose lives count, whose suffering matters, and who is rendered invisible.

What unites Carolina and the macaques is a shared position within these hierarchies. Both are made into instruments for the benefit of others; their agency is denied, and their capacities systematically stunted. Terms such as cruelty or abuse capture certain surface features of these cases, but they fail to illuminate the deeper mechanisms that reproduce such harms across time. What is required is a vocabulary capable of describing how social arrangements – rather than isolated acts – generate enduring patterns of injustice. In contemporary political theory, few thinkers have articulated this grammar of injustice more powerfully than Iris Marion Young.

2.1 Iris Marion Young and the Grammar of Oppression

Among her many contributions to political theory, Iris Marion Young (1990) developed one of the most influential contemporary accounts of oppression. She famously identified five faces of oppression – exploitation, marginalisation, powerlessness, cultural imperialism, and violence – thereby revealing its plural, structural, and systematic character. Her analysis also drew attention to three features that distinguish oppression from other moral wrongs: its objective nature, its grounding in social group relations, and the specific kind of harm it inflicts on those subjected to it.

Young's analysis is underpinned by a broader methodological reorientation in thinking about justice. She rejects the idea that justice should be approached primarily as a problem of distribution, as in John Rawls's *A Theory of Justice*. Instead, she argues that any adequate conception of justice must begin with domination and oppression – that is, with the social relations and institutional processes that produce and sustain inequality in the first place. Injustice, in this view, is not exhausted by unequal shares of goods but consists in the patterned organisation of social life that systematically advantages some while constraining others.

Closely related to this shift is Young's insistence that theories of justice remain attentive to concrete social contexts. Rather than deriving principles from idealised conditions, she calls for an inquiry grounded in the lived experiences of those who suffer injustice. Normative reflection, she argues, does not begin from an abstract moral standpoint but from within the social and political practices in which wrongs actually occur. Justice, on this view, is best approached not as an object of detached judgement but as a practice of attentiveness and response. "Normative reflection," Young writes, "arises from hearing a cry of suffering or distress, or feeling distress oneself"; the philosopher, she insists, "is always socially situated" (Young, 1990, p. 5). Justice is therefore rooted in concrete relationships and histories, not imposed from an abstract moral vantage point.

Young's emphasis on situatedness is nowhere clearer than in her account of oppression. The following discussion proceeds in two stages. First, it asks who can be the subject of oppression – that is, what kinds of beings can meaningfully be subjects of it. Second, it examines what distinguishes oppression from other forms of injustice. Together, these questions reconstruct Young's framework in a way that opens it to interspecies application.

2.2 Who Can Be Oppressed? Groups, Identity, and External Constitution

Let us begin by asking who can be the subject of oppression. Young's answer – widely accepted in the literature – is that oppression is directed at social groups rather than isolated individuals. This does not deny that individuals may suffer from these injustices; rather, it holds that their suffering arises in virtue of their membership in particular groups. What makes individuals vulnerable to oppression is not their individuality as such, but the social meanings and positions attached to the groups to which they are assigned.

Young defines a social group as "a collective of persons differentiated from at least one other group by cultural forms, practices, or way of life" (Young, 1990, p. 43).[1] What binds such a group together is not a set of shared attributes but a sense of identity or affinity that shapes patterns of association and differentiation. Groups, in this sense, are not fixed substances but relational constructs: "a group exists only in relation to at least one other group" (Young, 1990, p. 43).

This relational understanding of groups is crucial. Social groups are not defined primarily by internal traits but by patterns of social differentiation, exclusion, and mutual recognition. What constitutes a group is not a catalogue of shared properties but an act of identification occurring within a broader social field.

Importantly, this act of identification is not always voluntary or self-authored. While some groups are constituted internally through shared deliberation or solidarity, Young insists that group membership is often imposed rather than chosen. Individuals frequently find themselves thrown into social identities that precede them, in what Young, following Heidegger, calls "thrownness": "one *finds oneself* as member of a group, which one experiences as always already having been" (Young, 1990, p. 46).[2] They awaken to their group membership as something given, not something they have collectively willed.

This insight aligns Young's view with what I will call, following Ann Cudd, an externalist account of social identity (see below). For Young, "our identities are defined in relation to how others identify us, and they do so in terms of groups which are always already associated with specific attributes, stereotypes, and norms" (Young, 1990, p. 46). What matters, then, is not only how individuals see themselves, but how they are classified, treated, and positioned by others within existing structures of power. Such positioning need not presuppose reflexive self-understanding on the part of those positioned, it requires only systematic social treatment across institutional contexts.

A similar view is defended by Ann Cudd, who argues that group membership is determined not by internal psychological states but "by objective facts about the world, including how others perceive and behave toward that person" (Cudd, 2006, p. 36). In this externalist view, individuals may come to internalise or affirm the categories imposed upon them, but those categories originate in social processes of classification.

What the externalist account establishes is not that groups are always constituted from the outside, but that self-identification is not a necessary condition for group membership. Groups can come to exist through processes of external classification, exclusion, and social positioning, independently of whether the members of such groups recognise themselves as belonging to them. As Young observes, on occasions "a group comes to exist only because one group excludes and labels a category of persons, and those labeled come to understand themselves as group members only slowly, on the basis of their shared oppression" (Young, 1990, p. 46). Oppression is thus not always directed at groups that pre-exist it; it is often a process through which the very groups it targets are brought into being.

This point will prove crucial when we extend Young's framework beyond the human case. If group membership can be externally constituted – including through how beings are classified, exploited, and excluded by others – then self-identification or linguistic agency cannot be a necessary condition for being a subject of oppression. What matters, rather, is being positioned as a kind within social institutions and practices. As the sections that follow show, nonhuman animals are already collectively constructed in precisely this way: as subordinate beings whose identities are defined by use, inferiority, or disposability within human-dominated systems.

2.3 Oppression and the Obstruction of Self-Development

Not all injustices amount to oppression. While all oppression is unjust, not all injustice has the structural form or disabling effects that characterise oppression in Young's sense. To understand what makes oppression distinctive, we need to ask two questions: what kind of wrong it is, and what specific harm it entails.

Young treats oppression as a species of injustice rather than its synonym. The distinction turns on both the source and the form of the wrong. Oppression is not reducible to isolated acts of wrongdoing or to the misconduct of malicious individuals. It is a structural and institutional condition – a pattern of constraints embedded in the organisation of social life, sustained by laws, policies, norms, and everyday practices that collectively restrict the freedom and flourishing of certain groups. It is structural rather than episodic, persistent rather than occasional, and objective rather than merely experiential. Crucially, oppression need not be intentional: it can be sustained by ordinary, well-intentioned actions that reproduce unjust institutional arrangements without any individual agent intending to oppress. Nor does its existence depend on whether those who suffer it recognise their condition as such. One of oppression's most insidious effects is precisely its capacity to normalise itself, making injustice appear natural or inevitable to those it harms. Indeed, it is precisely because of this normalising effect that Young defends the structural and objective character of oppression: it provides, she argues, "a means of persuading others that a group is oppressed" – including when those who doubt it are the oppressed themselves (Young, 1990, p. 64).

Young also draws an important distinction between oppression and domination. Although both are forms of injustice, they target different dimensions of social life. Domination concerns restrictions on self-determination: the capacity of groups to participate meaningfully in collective decision-making and exercise control over the conditions of their own actions. Oppression, by contrast, targets the conditions of self-development – the material, social, and symbolic circumstances that enable individuals and groups to learn, communicate, cooperate, and cultivate their capacities. This distinction matters for the present argument. Whether and how nonhuman animals could be said to exercise collective self-determination – either directly or through representation – remains contested in animal political theory, and even the most developed defences of animal self-determination – most notably Donaldson and Kymlicka's sovereignty framework for wild animal communities[3] – operate at the *collective* level, leaving questions about the self-development of individual animals comparatively underdeveloped. The conditions of self-development, by contrast, are empirically evident across species and bear on individuals directly. Oppression therefore offers the more secure foothold for a structural account of what is done to animals in contemporary societies: not the denial of political authorship – a wrong whose application to animals remains philosophically contested – but the pervasive disabling and enclosure of their capacities to live and develop according to their own forms of life.

Young answers the question of what specific harm oppression entails by identifying it as fundamentally disabling. Oppression "consists in systematic institutional processes which prevent some people from learning and using satisfying or expansive skills in socially recognised settings, or which inhibit people's ability to play and communicate with others, or to express their feelings and perspective on social life in contexts where others can listen" (Young, 2000, p. 156). What oppression denies, then, is not liberty understood narrowly as freedom from interference, but the conditions of self-development. This idea lies at the core of Young's enabling conception of justice: justice concerns "the institutional conditions necessary for the development and exercise of individual capacities and collective communication and cooperation" (Young, 1990, p. 39). Injustice consists in social arrangements that obstruct or diminish those capacities, and oppression represents a particularly severe and pervasive form of such disablement.

This enabling conception marks a decisive break from distributive models of justice. Rather than focusing on the allocation of goods or rights, Young directs attention to the social conditions under which individuals and groups can become and act as agents. Justice, in this view, is not about equal shares but about enabling forms of life. As Rahel Jaeggi's non-essentialist account of self-realisation suggests, such development takes shape through practice rather than in abstraction (Jaeggi, 2014).

Yet Young's own account of self-development remains relatively underdeveloped, and its normative foundations have received surprisingly little systematic attention in the literature on structural injustice. A recent and important contribution to filling this gap is offered by Azizjon Bagadirov (2025), who argues that the concept of self-development requires both a clearer definition and a more precise account of the mechanisms through which structural oppression inhibits it. Drawing on John Dewey, Rahel Jaeggi, and Carol Gould, Bagadirov proposes a processual and non-essentialist account: self-development is best understood as the exercise and development of one's capacities in the pursuit of one's projects for their own sake. On this view, self-development is not a fixed end-state to be achieved, nor the realisation of some predetermined essence. It is a way of living – a mode of active engagement with the world, characterised by the continuous exercise of one's emotional, intellectual, physical, and other capacities in interaction with one's environment. What matters is not what one achieves but how one lives: whether one's life has the structure of ongoing development rather than stasis, constraint, or enforced passivity.

This processual account has two important implications for the chapter's argument. First, it is deliberately non-essentialist in a methodological sense: it does not require specifying in advance which capacities matter or what a flourishing life looks like for any particular kind of being. Capacities are identified inductively and contextually, without imposing a prior template drawn from any single form of life. This makes the account especially suited to interspecies application, where imposing a substantive conception of the good life risks reproducing precisely the anthropocentrism that a nonspeciesist theory of justice seeks to overcome. Second, and operating within the constraints set by this methodological commitment, the account grounds the normative weight of self-development in a developmental theory of well-being – associated in particular with Richard Kraut's work (Kraut, 2007) – according to which flourishing for any living thing consists in the development and exercise of its capacities as identified through that same contextual and inductive attention. This is not an assumption imported from outside but a generalisation arrived at by attending to what we already recognise as good: that a life characterised by the active exercise of one's capacities is, across a wide range of cases and kinds of beings, better than one in which those capacities are stunted, suppressed, or rendered inert.

Before this account can be extended to nonhuman animals, however, one further condition must be established. Self-development, as defined above, is not a purely biological process. It is a normatively significant one – it is what makes oppression a *wrong* rather than merely a fact. The framework applies to sentient beings: those capable of subjective experience, and therefore of having a welfare – a perspective from which things can go better or worse. Sentience is the ground of that welfare, and welfare is what gives the structural obstruction of self-development its normative weight. This condition also marks an important boundary. What distinguishes the structural obstruction of self-development from mere physical interference with a biological process is precisely that the former involves a being for whom things matter – a being with a perspective that can be foreclosed. Where sentience is absent, there is damage but not deprivation; interference but not injustice. It is worth noting that sentience alone is not sufficient: the framework also presupposes some minimal developmental capacity, a trajectory of becoming that oppressive conditions can obstruct or enclose. In practice, sentience and developmental capacity are rarely separable in the kinds of beings this account is most concerned with. Plants and non-sentient organisms may be damaged or destroyed by human action, and there may be independent reasons to care about this. But they are not subjects of oppression on the account developed here. Sentience is not a high bar – growing scientific consensus suggests it is met by a wide range of animals, including fish and many invertebrates – but it is a necessary one, and its role as a threshold condition should be kept in view throughout the analysis that follows.

Yet a developmental account of well-being, on its own, does not explain how oppressive structures systematically undermine flourishing from within. For that, we need a more fine-grained analysis of the mechanisms through which structural conditions shape not only what beings can do but what they can become. This is precisely what Bagadirov's concept of developmental enclosure provides. Building on feminist phenomenology and critical sociology, Bagadirov argues that structural oppression harms self-development not only externally – by blocking access to resources, opportunities, or social recognition – but internally, by shaping the embodied dispositions and perceptual horizons of those who inhabit oppressive conditions. Oppression, on this account, does not merely prevent certain actions; it structures the entire field within which development is possible, such that constrained, resigned, and enclosed dispositions become internalised as the normal conditions of existence. The oppressed do not simply face obstacles; they come to inhabit a world organised to foreclose certain possibilities from the inside.

Marilyn Frye's image of the birdcage captures part of this dynamic: oppression consists not in any single wire but in a network of constraints that, taken together, enclose and immobilise (Frye, 1983). Bagadirov deepens this image by showing that the cage not only constrains movement from the outside – it shapes the very dispositions through which the caged being relates to itself and its world. The harm of developmental enclosure therefore does not become fully visible at the level of individual acts or even of individual structural barriers.

In certain domesticated and captive animals, this process is not a theoretical inference but an empirically observable condition. Zoochosis – the range of stereotypic, repetitive behaviours displayed by animals in captivity, including incessant pacing, swaying, self-mutilation, and the compulsive repetition of apparently purposeless movements – is a direct expression of developmental enclosure. These behaviours are not natural to the animals that exhibit them; they emerge from conditions of chronic confinement that have structured the animals' embodied dispositions around the impossibility of self-development. The animal has not merely been prevented from doing certain things; it has been formed into a being whose behavioural and emotional repertoire has been systematically narrowed by the conditions of its existence. What zoochosis makes visible is the internal correlate of structural constraint: not the locked door, but the being that has been shaped by living behind it.

Selective breeding offers a further and, in some ways, more radical illustration. When dogs are bred over generations to amplify aggression for fighting, to suppress certain instincts in favour of docility, or to exaggerate physical traits – shortened muzzles, enlarged heads, extreme body proportions – that produce chronic pain, anxiety, or behavioural dysfunction, the enclosure of development does not operate through the conditions of any individual life. It is written into the biological inheritance of the animal before it is born. Here, developmental enclosure precedes the individual entirely: the possibilities for self-development available to a bulldog or a pit bull bred for fighting have been shaped by human purposes across generations, producing beings whose characteristic capacities have been distorted, diminished, or redirected in ways that serve human ends rather than canine flourishing. This is structural oppression operating at its deepest level – not merely constraining development within a life, but engineering the conditions of development before that life begins.

Together, zoochosis and the effects of selective breeding illustrate that developmental enclosure is not a metaphor but a measurable condition – one that produces observable, documented harms in the embodied lives of real animals. They also show that the concept applies with particular clarity to beings whose relationship with humans is most intimate and most thoroughly structured: domesticated and captive animals whose lives are, in the fullest sense, organised around human purposes rather than their own.

Extending this account from companion and captive animals to those caught up in industrial agriculture confirms its reach. Research in animal cognition, ethology, and comparative psychology consistently shows that social mammals, birds, and other cognitively complex animals develop embodied patterns of behaviour, emotional responsiveness, and social orientation through interaction with their environments – and that these patterns can be profoundly distorted by conditions of confinement, isolation, chronic stress, and the systematic denial of species-typical activities. A pig raised in a gestation crate does not merely lack the opportunity to root and explore; the conditions of her confinement shape her embodied dispositions in ways that make certain modes of being – curiosity, play, social navigation – structurally inaccessible rather than merely unrealised. A broiler chicken bred for rapid weight gain does not merely suffer from physical pain; the engineering of her body forecloses the very capacities whose exercise would constitute development. These are not simply cases of frustrated preference or blocked opportunity. They are further instances of developmental enclosure: the structural production of beings whose possibilities for self-development have been shaped, by the very conditions of their existence, into the form that oppression requires.

This concept adds a crucial dimension to Young's framework. Where Young's five faces of oppression describe the external mechanisms through which self-development is inhibited, developmental enclosure describes the internal correlate: the way those mechanisms come to inhabit the beings they oppress. Together, they provide a more complete account of the harm of structural oppression – one that captures not only what animals are prevented from doing but what they are prevented from becoming.

In summary, oppression in Young's account is marked by several interconnected features. First, it is structural: it arises from the patterned organisation of social life rather than from isolated acts of wrongdoing. Second, it is independent of the intentions of those who perpetuate it: structural oppression can be sustained by ordinary, well-intentioned actions that reproduce unjust institutional arrangements without any individual agent intending to oppress. Third, it is objective rather than subjective: its existence does not depend on whether those who suffer it recognise their condition as oppressive. Fourth, it is disabling: it systematically impedes the development and exercise of capacities necessary for agency and participation. To these we can now add a fifth dimension, contributed by Bagadirov's analysis: oppression is also formative – it shapes the embodied dispositions and developmental horizons of those it targets, producing beings whose possibilities for self-development have been enclosed rather than merely obstructed from outside. And undergirding all five features is the sentience threshold: the framework applies to beings for whom these conditions are experienced, not merely undergone.

Together, these features distinguish oppression as a particularly pernicious form of injustice – and, crucially, they provide the conceptual tools required to extend this analysis beyond the human case. The second and third features are especially significant for that extension: if oppression requires neither malicious intent on the part of perpetrators nor conscious awareness on the part of those who suffer it, then the most common grounds for denying that animals can be oppressed dissolve. To speak of animal oppression is therefore not to engage in metaphor but to name a real and serious form of structural injustice – one that disables and encloses nonhuman animals' capacities to develop and act on their own terms.

3 Animals and Oppression

Earlier in this chapter, the case of the macaques abducted from the monastery was offered as an example of animal oppression. Applying the vocabulary of oppression to animals is not common practice, however, and many remain sceptical about extending this normative language beyond the human sphere. Even those sympathetic to animal welfare or animal rights may hesitate to adopt the political vocabulary of oppression – agreeing that animals are routinely harmed through human practices, yet remaining unsure whether those harms amount to injustice rather than simply moral wrong or tragedy. Because one of the central aims of this book is to show that animals' suffering must be understood not only in moral but in political terms – as the result of structural injustices rather than isolated cruelties – it is necessary to face these objections directly.

3.1 Are Animals Aware of Their Oppression?

This first objection assumes that subjects of oppression must be aware of their own condition – that they must understand themselves as victims of structural injustice. On this view, if those who are subordinated cannot conceptualise their own subordination, then they may be harmed, but the concept of oppression does not apply.

Yet this argument rests on a misconception of what oppression requires. Oppression is not a subjective condition but an objective social relation, and crucially, its existence does not depend on whether those who suffer it recognise their condition as such. Indeed, one of oppression's most insidious effects is its capacity to normalise injustice to the point that it becomes invisible, not only to those who benefit from it but also to those who endure it.

This becomes clear when we distinguish oppression from harm. A slave may resent the master's individual acts of cruelty – the whippings, punishments, and daily humiliations. These are indeed personal harms. But what makes slavery a paradigmatic case of oppression is not these acts alone; it is the underlying structure that defines the slave as property and legitimises the master's power. Oppression refers not to the acts but to the social order that sustains and normalises them. The slave need not conceptualise this structure for it to govern their life. For example, the slave may be convinced that their terrible life is the result of their master's awfulness rather than the outcome of oppression. Yet that belief does not change the fact that they are oppressed. The same holds for animals.

History provides many other examples. Under patriarchal or colonial systems, the oppressed have often internalised their subordination, accepting their position as natural or deserved. To suggest that those who internalised their subordination were therefore not oppressed would be to claim that oppression ceases to exist precisely when it succeeds most completely. This is untenable. As Young notes in her discussion of Jews in Vichy France, many assimilated Jews had no specifically Jewish identity until they were marked as such by others and assigned a distinct social status. These people, in Young's phrase "'discovered' themselves as Jews," and from that imposed classification formed a group identity and affinity with one another (Young, 1990, p. 46). The example illustrates that group membership – and the oppression that targets it – can be externally constituted, taking effect prior to and independently of self-identification.

The same point holds for animals. The criterion for oppression lies not in the mental states of its victims but in their structural position – in being situated within a hierarchy that systematically constrains the conditions for self-development. On this account, the question of whether animals can conceptualise their own subordination does not bear on whether they are subjects of oppression.

3.2 Do Animals Belong to an Oppressed Group?

A second objection targets another of Young's defining features of oppression: that it is directed not at isolated individuals but at groups. On this basis, sceptics argue that animals do not form a relevant group at all. The term encompasses an immense range of living beings – mammals and birds, insects and fish, domesticated and wild creatures – too heterogeneous to count as a single oppressed category. What could a broiler chicken and a laboratory macaque, a companion dog and a tuna, possibly share that would make them members of the same politically relevant group?

At first glance, this objection carries some weight. The biological diversity encompassed by the category *animal* is indeed far greater than that within human groups typically identified as oppressed. But this is precisely not what makes a group politically relevant. As argued in Section 2.2, Young's account treats social groups as defined by social relations of identification and differentiation, not by shared natural or biological traits. What matters is not what members share by nature but how they are collectively perceived, classified, and positioned within social structures.

This relational account of group formation applies directly to animals. In industrialised societies, nonhuman animals are collectively positioned within a human–animal hierarchy through a set of socially constructed markers that operate across all the biological diversity the objection invokes. The criteria used to constitute animals as a subordinate group – lack of reason, absence of language, perceived irrationality, legal status as property – are not descriptions of natural facts but acts of social classification that organise the group's position within existing hierarchies of power and value. What unites a broiler chicken and a laboratory rat as members of the same politically relevant group is not their biological similarity but their shared structural position: both are legally classified as property, both are excluded from the protections that attach to persons, and both are available for use, exploitation, and disposal by human institutions that treat their interests as subordinate to human purposes.

The law is perhaps the most powerful evidence for this claim. Property law, animal welfare legislation, and food safety regulation all treat the category *animal* as a unified legal grouping, regardless of the enormous biological differences between the beings it encompasses. This legal unity is not a reflection of natural homogeneity; it is the institutional expression of a social hierarchy. The law constitutes animals as a group precisely by assigning them a shared subordinate status that determines what can be done to them, who has authority over them, and what protections, if any, they are entitled to claim. The group is not discovered in nature; it is produced through the institutional arrangements that sustain it.

Val Plumwood famously described this process as othering – the systematic construction of the nonhuman as the inferior counterpart to the human (Plumwood, 1993). This logic of dualism treats the human as the standard of value and the animal as its negation: irrational, instinctual, voiceless, and available for use. Through such oppositions, animals become a politically real group, not because of intrinsic unity but because of how they are collectively defined and treated by human institutions. The pattern is structurally identical to the one analysed in Section 2.2: groups can come into being through the very practices of exclusion, classification, and subordination that target them (Young, 1990, p. 46). Animals are no exception. They are constituted as a group through precisely these practices.

A residual worry remains. Even if animals can be said to constitute a politically relevant group, the application of the self-development framework might seem to exclude animals at the margins of cognitive complexity – insects, fish, and others whose capacities for development are uncertain or contested. As established in Section 2.3, however, the framework's sentience threshold addresses this worry directly. The account of self-development is non-essentialist: it refers to the exercise and development of whatever capacities a being possesses in pursuit of its own modes of engagement with the world. What the threshold requires is that the being in question be sentient – capable of subjective experience, and therefore of having a welfare that can be benefited or harmed. Growing scientific consensus suggests this condition is met by a wide range of animals, including fish and many invertebrates. Beings that do not meet this threshold – plants, non-sentient organisms – are not subjects of oppression on this account, not because they do not matter in other respects, but because oppression specifically requires a being for whom structural foreclosure is *suffered* rather than merely undergone.

In this light, the category *animals* fulfils the criteria required by Young's analysis. It is socially constructed, relationally defined, institutionally enforced, and hierarchically positioned relative to a privileged group. The biological diversity of the category is not an obstacle to its political relevance; it is, if anything, further evidence of how thoroughly the human-animal hierarchy operates – organising beings of radically different kinds into a single subordinate class through the shared machinery of property law, economic exploitation, and cultural devaluation.

To speak of animal oppression is therefore not to stretch Young's concept beyond its limits. It is to apply it consistently, recognising that the same mechanisms that construct racial, gendered, and class-based hierarchies also construct and maintain species hierarchies. In all these cases, the oppressed are defined through their assigned difference from the dominant group and systematically denied the conditions under which their capacities for self-development can be expressed and cultivated.

3.3 Does Animal Oppression Trivialise Human Oppression?

A third objection comes not from those who are indifferent to animal suffering but from those who are deeply committed to the analysis and contestation of human oppression – and it is, in some respects, the most serious version of the zero-sum objection introduced in Chapter 1. The worry, in its most considered form, is this: by extending the concept of oppression to nonhuman animals, we risk equalising or even diminishing the distinctive historical weight of human cases.[4]

The oppression of enslaved Africans, of colonised peoples, of women under patriarchy – these are not abstract structural phenomena but lived histories of specific violence, resistance, memory, and struggle. To place them in the same conceptual category as the treatment of factory-farmed animals risks flattening these histories, suggesting a moral equivalence that dishonours those who suffered and those who continue to suffer the legacies of these injustices.

This objection deserves to be taken seriously on its own terms, and it should not be dismissed as mere anthropocentrism. It reflects a legitimate concern about the political uses of analogical reasoning. Comparisons between human and animal oppression have historically been deployed in ways that dehumanise rather than liberate – used to animalise racialised, colonised, or disabled people as a technique of degradation rather than as a gesture of solidarity. The history of such comparisons is not neutral, and any extension of the oppression framework to animals must reckon honestly with that history.

What this objection conflates, however, is structural resonance with moral identity. This book does not claim that the oppression of animals is the same as the oppression of enslaved people, or of colonised peoples, or of women. It does not claim that these histories are interchangeable, that their harms are equivalent in scale or kind, or that the political and legal responses they demand are identical. The histories of human oppression are specific, irreducible, and carry moral and political weight that cannot be absorbed into a general theory without remainder.

What the book does claim is something more modest and more precise: that these instances of injustice share a similar grammar – a similar underlying logic through which beings are sorted into hierarchies of value, their capacities enclosed, their subordination normalised through institutional arrangements that operate independently of individual intent. Young's five faces of oppression describe a structural pattern that recurs across different historical contexts and different kinds of beings. Identifying that pattern in the treatment of animals does not erase the differences between cases; it illuminates the architecture of injustice that makes each case possible.

This distinction between structural resonance and moral identity is not a concession but a methodological requirement. If the concept of oppression is to do genuine analytical work – if it is to be more than a rhetorical intensifier applied to whatever wrongs we find sufficiently serious – then it must be capable of identifying shared structural features across different cases without collapsing those cases into one another. That is precisely what Young's framework is designed to do. As she argued, oppression is not a single thing but a family of structural injustices, whose specific forms and histories vary enormously while sharing a common underlying logic. Extending that framework to animals is therefore not a departure from its analytical ambitions but a consistent application of them.

Some of the most important recent work on the relationship between human and animal oppression has argued not that these oppressions are identical but that they are structurally entangled: that the logic of dehumanisation that underwrites racism and colonialism and the logic of animalisation that underwrites speciesism are not separate systems but mutually reinforcing ones (Kim, 2015; Ko and Ko, 2017; Boisseron, 2018). On this view, taking animal oppression seriously is not a distraction from anti-racist or anti-colonial politics but a condition of their full critical force.

To identify a shared grammar of oppression is not to claim that all instances of injustice are the same. It is to insist that the structural logic through which some lives are rendered disposable, some capacities are systematically enclosed, and some forms of existence are organised around the benefit of others – that this logic is not the exclusive property of any single history. Naming it across the species boundary does not diminish the human cases it has been used to describe. It extends the demand for justice to those who have been excluded from it.

4 The Faces of Animal Oppression

Even if animals can, in principle, be victims of oppression, one still needs to establish whether they are, in fact, oppressed. To answer this, we must look beyond isolated harms or acts of cruelty to the broader structures that organise human-animal relations. Here, Iris Marion Young's five faces of oppression – exploitation, marginalisation, powerlessness, cultural imperialism, and violence – offer a powerful analytic lens.

Young developed these categories to capture the plural and intersecting mechanisms through which social groups are systematically disadvantaged. Each face marks a distinct form of structural harm, but all share a common feature: they obstruct the conditions of self-development.

Once we acknowledge that animals form a socially constituted group situated at the bottom of a human–animal hierarchy, Young's framework provides the conceptual tools to trace the many ways oppression operates across species. Animals are not merely harmed or used – they are structurally positioned within institutional systems that constrain their agency and foreclose their possibilities for development. This argument was first advanced by Lori Gruen (2009), whose pioneering application of Young's framework to human–animal relations opened a line of inquiry subsequently developed by Robert Jones (2015) and Mathieu Dubeau (2019). The analysis that follows builds on and extends these contributions, moving through each of Young's five faces of oppression to show how they map onto the institutional arrangements that sustain human–animal hierarchy.

4.1 Exploitation

Young defines exploitation as a structural process through which "the energies of the have-nots are continuously expanded to maintain and augment the power, status, and wealth of the haves" (Young, 1990, p. 50). It is not an occasional act of injustice but a structural relation between social groups, sustained by social rules about who works, who benefits, and how the results of labour are appropriated.

In the case of animals, this process is literal. As Gruen observes, the "lives and bodies of female animals on factory farms are completely controlled to produce the maximum amount of product at the smallest cost to the producer [...] and when their labour is thoroughly extracted, they are killed" (Gruen, 2009, p. 162).

Exploitation encompasses both the use of labour and the consumption of life. Animals' biological and reproductive capacities – gestation, lactation, movement, and growth – are commodified in systems that treat their suffering as a cost of doing business. Profit maximisation becomes the organising principle, and the suffering of sentient beings is converted into an accounting variable.

The injustice lies not only in the pain inflicted but in the very organisation of production that turns animals into instruments. Exploitation thus operates as the economic foundation of interspecies oppression.

4.2 Marginalisation

Marginalisation, which Young considers "perhaps the most dangerous face of oppression," involves systematic exclusion from participation in social life. It is not merely material deprivation but a denial of recognition and purpose – such that even if marginalised beings were provided a comfortable material life, "injustices of marginality would remain in the form of uselessness, boredom, and lack of self-respect" (Young, 1990, p. 55).

Animals are profoundly marginalised in this sense. They are excluded from their own forms of life, their capacities constrained or rendered irrelevant. In farms and laboratories, they are prevented from expressing their natural behaviours and reduced to functions of production. As Jones notes, marginalisation occurs when animals are routinely forced to live and die "in conditions under which they are deprived the ability to express requisite species-typical behaviours and capacities" (Jones, 2015, p. 475).

The most extreme expression of this exclusion is disposability. In industrial farming, male chicks and calves – deemed unprofitable – are killed by suffocation, gassing, or crushing shortly after birth. These are not aberrations but institutionalised practices. In Scotland, the Farm Advisory Service explicitly recommends culling unproductive cows out of the system (FAS, 2020). This is marginalisation codified into policy. Animals' lives are stripped of intrinsic value and reduced to economic variables. Their existence becomes conditional on productivity, and their deaths are predictable outcomes of systemic design.

Mathieu Dubeau has argued that disposability is a defining feature of animal marginalisation, particularly in the case of companion animals, who may be abandoned or euthanised when deemed unwanted or deficient (see Dubeau, 2019, p. 611). While this insight captures a genuine dimension of animal vulnerability, it understates the structural and institutional character that makes marginalisation a form of oppression. Companion animals, though on occasions abandoned, are often protected by law and by social norms that grant them at least some degree of moral standing. By contrast, the disposability of farmed or laboratory animals is systematically produced and normalised through institutional practices, legal frameworks, and economic imperatives. Marginalisation qualifies as oppression most clearly where disposability is not the exception but the rule – where it is built into the design of the system itself.

4.3 Powerlessness

The third face, powerlessness, captures the structural asymmetry of agency between dominant and subordinate groups. The powerless are "those over whom power is exercised without their exercising it" (Young, 1990, p. 56). They are denied the ability to make decisions, shape outcomes, or even refuse participation.

This condition defines the lives of most animals under human control. As Jones explains, powerlessness "involves inhibition in the development of an animal's autonomy and capacities, and an exposure to disrespectful treatment because of the subordinate status that the animal occupies" (Jones, 2015, p. 474). The institutions in which this powerlessness is enacted span the entire infrastructure of contemporary human–animal relations: as Gruen observes, from "zoos to feedlots, pet shops to laboratories, factory to fur farms, nonhuman animals are denied the most basic control over their lives" (Gruen, 2009, p. 163). They do not choose when to eat, where to move, or with whom to live. Their reproductive lives, social bonds, and bodily integrity are subject to human command. This institutional subordination, enshrined in property law and justified by cultural norms, reduces animals to perpetual dependents, structurally incapable of exercising the agency that Young associates with self-development.

Powerlessness is therefore not simply a lack of power but a condition of structural dependency. It renders animals passive within their own subordination, ensuring that oppression reproduces itself as the normal condition of their existence.

4.4 Cultural Imperialism

Cultural imperialism, Young's fourth face, refers to "the universalisation of a dominant group's experience and culture, and its establishment as the norm" (Young, 1990, p. 59). In this process, the dominant group's worldview becomes the measure of all value, while the perspectives of others are rendered invisible or inferior.

For animals, this takes the form of a pervasive anthropocentrism that treats human norms – rationality, productivity, property – as universal benchmarks of value. This is, in Val Plumwood's language, the logic of othering: the construction of animals as deficient beings whose existence is meaningful only in relation to human purposes (Plumwood, 1993). Adapting Young's formulation to the species case, Jones (2015, p. 474) characterises this as the universalisation of a dominant species' experience and culture, against which other species are measured and devalued.

This logic manifests across a vast array of practices: in entertainment industries that stage animals as spectacles; in conservation schemes that value animals primarily for their contribution to environmental goals rather than for their own individual forms of life; and in the domestication of companion animals, whose lives are scripted around human schedules, spaces, and expectations. Animals' own forms of communication, community, and joy are subordinated to human interpretation and rendered legible only through a human lens.

This critique is not, however, evenly applicable across all cultural contexts. Both Gruen (2009) and Jones (2015) discuss such cases – Gruen the Makah whale hunt, Jones the Gadhimai animal sacrifice festival in Nepal – as examples of cultural imperialism exercised across species. Yet, as Gruen rightly recognises, condemning these practices in isolation risks reproducing another form of imperialism: imposing dominant Western moral frameworks upon already marginalised human communities. In many such cases, oppression is layered and intersectional – Indigenous or colonised peoples may simultaneously resist human imperialism while sustaining practices that involve animal subordination. This tension is real and should not be dissolved by easy analogy.

The framework developed in this book does not claim universal jurisdiction over culturally embedded practices. As Chapter 7 argues in the context of food sovereignty, the institutional principles of care and solidarity are intended to govern state and institutional design in the political communities this book primarily addresses – not to serve as a global template that overrides the claims of communities whose relationship to animals is embedded in different histories, ecologies, and forms of life. Acknowledging this limit is not a retreat from the book's central argument; it is a recognition of where the argument has force and where it does not.

Cultural imperialism thus naturalises species hierarchy: it is the institutional process through which anthropocentric standards are made to appear self-evident, erasing animals' perspectives while framing oppression as care, stewardship, or necessity. The result is a political arrangement that appears, instead, to be a fact of nature.

4.5 Violence

The final face of oppression, violence, exposes the raw coercive force that sustains the entire system of structural injustice. Young identifies it as both physical and symbolic, ranging from direct attacks to the constant threat that defines the existence of oppressed groups, a persistent possibility on the horizon of the social imagination (Young, 1990).

For billions of animals, violence is not a possibility hovering on the horizon but a certainty inscribed in the conditions of their existence. From slaughterhouses to laboratories, they are confined, mutilated, and killed in ways that are lawful, routine, and socially sanctioned – not as exceptions to the normal order but as expressions of it. As Gruen observes, the food industry is "violent from start to gruesome finish" (Gruen, 2009, p. 165).

This is not random cruelty but structural violence: the organised and legitimised infliction of harm in pursuit of human ends. It is analogous to the use of pesticides in agriculture – deliberate acts of extermination justified by productivity. Just as unwanted insects are killed to protect crops, elephants and other animals are destroyed when they interfere with profit. In both cases, violence functions as a management tool, embedded in systems that treat death as a necessary cost of production.

What makes this violence oppressive is not the excess of the system but its normal functioning – through laws that authorise it, markets that require it, and consumption habits that depend on it remaining invisible. The suffering of animals is not a scandal to be exposed but a condition to be managed; their deaths are not tragedies but transactions. This is what structural violence looks like when it has succeeded completely: not exceptional, but ordinary.

Taken together, these five faces reveal the architecture of interspecies oppression. Exploitation drains animals' capacities for human benefit; marginalisation excludes them from participation in their own forms of life; powerlessness silences their agency; cultural imperialism reduces their lives to the experiences and interpretations of others; and violence enforces the hierarchy that binds them all. Each face sustains the others: exploitation relies on powerlessness, cultural imperialism legitimises marginalisation, and violence maintains both. The result is a coherent and self-reproducing structure that denies animals the very conditions of self-development that Young identifies as central to justice – and does so not only by blocking their capacities from the outside but, as Bagadirov's account of developmental enclosure makes clear, by forming the very beings those capacities belong to.

To name these injustices as oppression is not a rhetorical exaggeration. Oppression is a more specific concept than injustice: while injustice can cover a wide range of wrongs – maldistribution, procedural failures, individual harms – oppression names a wrong that is irreducibly relational and structural. It presupposes a group hierarchy, a dominant and a subordinate, and a set of arrangements that actively reproduce that hierarchy across time. To say that animals are oppressed is not merely to say that they suffer, or that their interests are insufficiently weighted in moral calculation. It is to say that they occupy a subordinate position within a social order. This distinction also bears on how we think about responses. If the problem were injustice in the generic sense, correction or reform might suffice – adjusting welfare standards, extending legal protections. But if the problem is oppression, the appropriate response is resistance: challenging the hierarchies that produce subordination and dismantling the arrangements that normalise it. What forms of political action this resistance can legitimately take is a question the book returns to in its final part.

5 Conclusion

This chapter opened with two stories. The first concerned Carolina, a 15-year-old Guatemalan child, packing cereal boxes in an American factory in the middle of the night, very far from home. The second concerned a group of rhesus macaques, smuggled from a monastery in northern Thailand, transported through falsified supply chains, and left in American biomedical laboratories to live and die in sterile cages. These stories were offered as windows onto a shared structural condition – one in which human and nonhuman lives are organised within systems that convert bodies into instruments, existence into economic value, and vulnerability into a resource to be exploited.

The chapter's central argument has been that what Carolina and the macaques share is not merely suffering but oppression: a specific and serious form of injustice defined by its structural character and its disabling effects. Drawing on Iris Marion Young's framework, the chapter identified five core features: that oppression arises from entrenched group hierarchies rather than isolated wrongdoing; that it need not be perpetuated with malicious intent; that its existence does not depend on whether those who suffer it recognise their condition as such; that it is fundamentally disabling, systematically obstructing the conditions of self-development; and that it is formative – shaping the embodied dispositions and developmental horizons of those it targets, enclosing rather than merely obstructing their possibilities from outside. Sentience was established as the threshold condition: oppression, as a structural wrong, requires a being for whom developmental foreclosure is experienced rather than merely undergone.

Three objections to extending oppression beyond the human case were addressed. The first – that animals cannot conceptualise their own condition – fails because oppression is an objective structural relation, not a subjective experience. The second – that animals do not constitute a relevant group – fails because group membership is constituted by shared structural position within socially produced hierarchies, not by biological homogeneity or self-identification. The third, more searching objection holds that extending oppression to animals risks diminishing the distinctive historical weight of human cases. The chapter argued that this conflates structural resonance with moral identity. What is claimed here is not that animal oppression is the same as the oppression of enslaved or colonised peoples. It is that these instances of injustice share a similar grammar – a logic through which beings are sorted into hierarchies of value, their capacities enclosed, and their subordination normalised. Identifying that grammar across the species boundary does not flatten the differences between cases. It extends the demand for justice to those who have been excluded from it.

Tracing Young's five faces of oppression showed how animals in industrialised societies are systematically positioned at the bottom of a human-animal hierarchy whose architecture mirrors, and in important respects reinforces, the hierarchies that organise human oppression. These faces do not operate independently: exploitation relies on powerlessness; cultural imperialism legitimises marginalisation; violence maintains both. The result is a coherent and self-reproducing structure that denies animals the conditions of self-development – and does so not only by blocking their capacities from the outside but by forming the very beings those capacities belong to.

The oppression of animals is not an anomaly within an otherwise just order but a constitutive feature of the institutional arrangements of industrialised societies, woven into property law, economic systems, cultural norms, and political classifications in ways that make it structurally continuous with, rather than merely analogous to, the oppression of human groups. Recognising this requires not only a revision of moral concern but a transformation of political imagination: the structures that produce animal oppression must be analysed, contested, and dismantled with the same seriousness that political theory has brought to human injustice.

The chapters that follow extend this diagnostic framework across three further sites of structural injustice. Chapter 3 examines the entanglement of ableism, capitalism, and speciesism, showing how the same logic of productive value that renders disabled humans disposable also organises the exploitation and culling of animals in industrial agriculture. Chapter 4 turns to migration and borders, analysing how the same infrastructures that criminalise human migrants simultaneously fragment ecosystems and generate structural vulnerability for nonhuman animals. Chapter 5 examines global extractivism, showing how industries such as palm oil production simultaneously displace Indigenous communities and destroy the habitats of animals whose world is consumed by the same commodity chains. Taken together, these chapters demonstrate that the grammar of oppression identified here is not confined to any single domain or species – it is a recurring and constitutive pattern in the organisation of social life, manifesting across different spheres through distinct mechanisms yet according to a shared and identifiable underlying logic.

Notes

1 Young distinguishes social groups from two other forms of collectivity: aggregates and associations. Aggregates are statistical groupings – such as people classified by income or height – that lack shared interaction, identity, or social meaning. Associations, by contrast, are voluntary organisations formed for specific purposes, such as trade unions or political parties. Social groups differ from both in that individuals are not conceptually prior to them; rather, persons are constituted, in part, through their group affiliations. As Young puts it, "A person's particular sense of history, affinity, and separateness, even the person's mode of reasoning, evaluating, and expressing feeling, are constituted partly by her or his group affinities" (Young, 1990, p. 45). The qualifier "*partly*" is crucial, as it preserves space for individual agency and responsibility while acknowledging the formative power of social belonging.

2 On the original concept, see Heidegger, 1962, §29. See also Withy, 2014.

3 Donaldson and Kymlicka, 2011. For critical engagement see, among others, Horta, 2013; Cochrane 2013; Ladwig, 2015.

4 I am grateful to Rosine Kelz for pressing me on this important issue.

References

Bagadirov, A. (2025). Structural injustice and self-development. *Journal of Social Philosophy*. doi:10.1111/josp.70016.

Boisseron, B. (2018) *Afro-dog: Blackness and the animal question.* Columbia University Press.

Cochrane, A. (2013) Cosmozoopolis: the case against group-differentiated animal rights. *Law Ethics and Philosophy*, 1, 127–141.

Cudd, A. (2006). *Analyzing oppression.* Oxford University Press.

Donaldson, S., & Kymlicka, W. (2011) *Zoopolis. A political theory of animal rights.* Oxford University Press.

Dreier, H. (2023, February 25). Alone and exploited, migrant children work brutal jobs across the U.S. *The New York Times.* www.nytimes.com/2023/02/25/us/politics/migrant-child-labor-exploitation.html.

Dubeau, M. (2019). Species-being for whom? The five faces of interspecies oppression. *Contemporary Political Theory*, 19, 596–620. doi:10.1057/s41296-019-00363-7.

FAS/Farm Advisory Service. (2020, May 6). Maximising the power of culling as a management tool. www.fas.scot/article/maximising-the-power-of-culling-as-a-management-tool/.

Frye, M. (1983). *The politics of reality: Essays in feminist theory.* Crossing Press.

Gruen, L. (2009). The faces of animal oppression. In A. Ferguson & M. Nagel (Eds.), *Dancing with Iris: The philosophy of Iris Marion Young* (pp. 161–172). Oxford University Press.

Heidegger, M. (1962). *Being and time* (J. Macquarrie & E. Robinson, Trans.). Harper & Row. (Original work published 1927)

Horta, O. (2013) Zoopolis, intervention, and the state of nature. *Law, Ethics and Philosophy*, 1, 113–125.

Jaeggi, R. (2014). *Alienation* (F. Neuhouser & A. Smith, Trans.). Columbia University Press.

Jones, R. (2015). Animal rights is a social justice issue. *Contemporary Justice Review*, 18(4), 467–482. doi:10.1080/10282580.2015.1092960.

Kim, C. J. (2015) *Dangerous crossings: Race, species, and nature in a multicultural age.* Cambridge University Press.

Ko, A., & Ko, S. (2017). *Aphro-ism: Essays on pop culture, feminism, and Black veganism from two sisters.* Lantern Books.

Kraut, R. (2007). *What is good and why: The ethics of well-being.* Harvard University Press.

Ladwig, B. (2015) Against wild animal sovereignty: an interest-based critique of Zoopolis. *Journal of Political Philosophy*, 23, 282–301.

Plumwood, V. (1993). *Feminism and the mastery of nature.* Routledge.

US Department of Justice. (2022). Cambodian officials and six co-conspirators indicted for primate smuggling scheme. www.justice.gov/usao-sdfl/pr/cambodian-officials-and-six-co-conspirators-indicted-taking-part-primate-smuggling-0.

Withy, K. (2014). Situation and limitation: Making sense of Heidegger on thrownness. *European Journal of Philosophy*, 22, 61–81. doi:10.1111/j.1468-0378.2011.00471.x.

Young, I. M. (1990). *Justice and the politics of difference.* Princeton University Press.

Young, I. M. (2000). *Inclusion and democracy.* Oxford University Press.

3

DISABLED BY DESIGN

Ableism, Capitalism, and Animal Oppression

1 Introduction

In late June 2025, British journalist Patrick Christys, speaking on GB News with American commentator Lewis Schaffer, discussed the alleged need to curtail welfare provision in the United Kingdom. When Christys asked how disability benefits might be eliminated – how one might get people with disabilities off their backsides – Schaffer replied: "Just starve them. That's what people have to do. That's what you've got to do to people. You can't just give them money. What else can you do? Shoot them? I mean, I'd suggest that, but I think that's maybe a bit strong" (Pring, 2025).

At first glance, this exchange might be dismissed as an instance of individual ignorance or provocation. Yet to do so would be to misrecognise its significance. Rather than an isolated lapse, such rhetoric is better understood as symptomatic of deeper social and economic structures through which contemporary Western societies conceptualise disability, dependency, and value. This chapter argues that discourses like Schaffer's are not aberrational but expressive of a broader political economy in which the marginalisation of people with disabilities is not accidental but constitutive of how social life is organised under capitalism.

Central to this organisation are evaluative norms that measure lives according to productivity, efficiency, and economic contribution. These norms are not confined to the human sphere. From their inception, capitalist regimes of valuation have operated across species boundaries, sorting bodies and capacities into hierarchies of usefulness and expendability. Nowhere is this logic more visible than in industrial animal agriculture, a paradigmatic site of structural oppression. Within this system, animals are subjected to a relentless calculus of productivity: those who meet imposed efficiency thresholds are

DOI: 10.4324/9781003434856-4

normalised and instrumentalised, while those who fail – because of injury, illness, or nonconforming embodiment – are systematically culled and rendered waste.

This same logic operates in the marginalisation of people with disabilities, whose lives are similarly devalued when their embodied capacities diverge from dominant norms of efficiency and output. The chapter therefore advances a dual claim. First, ableism and the enforcement of norms of normalcy function as central mechanisms in the oppression of people with disabilities. Second, these same mechanisms play a constitutive role in sustaining the systematic subjugation of nonhuman animals. This second claim builds on Sunaura Taylor's pathbreaking analysis in *Beasts of Burden* (2017), which articulated the structural entanglement of ableism and animal oppression. The chapter develops this connection in a specifically political-economic register, drawing on Young's structural injustice framework (Young, 2011) and analysing contemporary biotechnological innovations in industrial poultry production that have emerged since Taylor's book.

Drawing on disability theory – including Campbell (2009), Garland-Thomson (2002), Barnes (2016), Taylor (2017), and Jenkins and Webster (2020) – the chapter shows how practices of oppression are co-constituted across species lines through shared constructions of embodiment, productivity, and value. In continuity with the framework developed in the previous chapter, oppression is understood here as a structure of irresponsible privilege: an arrangement that enables certain lives to develop and flourish while rendering others disposable, exploitable, or eliminable.

The argument is developed through an analysis of the contemporary poultry industry, where capitalist and ableist logics intersect with particular clarity. Practices such as the selective breeding of chickens for maximum productivity, the pre-hatch elimination of male embryos, and the development of featherless birds designed to reduce production costs illustrate how technologies of optimisation are mobilised not to support animal flourishing but to regulate vulnerability, enforce norms of bodily conformity, and justify large-scale elimination. These cases reveal how ableism operates as a multispecies technology of governance – one that organises life itself according to standards of worth that systematically exclude the vulnerable.

2 What Is Disability?

When considering what counts as disability, what immediately stands out is the extraordinary diversity of conditions encompassed by the term. According to the World Health Organization (World Health Organization & World Bank, 2011), approximately 1.3 billion people worldwide – around 16 per cent of the global population – live with a disability. This category includes individuals who have lost a limb or experience difficulty walking; those with Down syndrome or Attention Deficit Hyperactivity Disorder; people who are deaf, live with depression, or experience learning differences such as dyslexia or dyscalculia. While all of these individuals may be classified as disabled, their circumstances, needs, and lived experiences vary dramatically across physical, sensory, cognitive, and psychosocial dimensions, and are further shaped by intersecting social positions such as class, race, gender, and access to care. This radical heterogeneity raises a foundational question: what, if anything, unifies such disparate conditions under the category of disability?

One influential response is the naturalistic model. In this view, disability is understood as a set of objectively identifiable characteristics that can be empirically observed and measured in the body or mind. Among its most prominent proponents is Christopher Boorse (1975, 1977, 1997), who defines disability in terms of dysfunctions or anomalies identified through a statistical analysis of bodily and mental functions. Central to Boorse's account is the notion of species-typical function, conceived as a descriptive, value-neutral account of how an organism's parts normally function. Disabilities, in this view, are explained by anomalous states – anomalous insofar as they are statistically atypical and detrimental to the individual's survival.

Without rehearsing all the objections to naturalistic accounts in detail,[1] it is worth noting Derek Bolton's central criticism: Boorse's account, which treats "mere statistical difference from some population norm" as constitutive of disease or its mental equivalent, pathologises ordinary human variation (Bolton, 2008, p. 115). The political consequences are clear. By construing disability as deviation from a biological norm, the naturalistic approach marginalises entire groups of people and dismisses the self-understanding of those who now contest such classifications. More importantly for this chapter, it obscures the social, political, and economic conditions that shape how impairments are lived, interpreted, and rendered disabling.

The social model of disability addresses precisely this gap. Rather than locating disability in bodies and minds alone, it shifts attention to the wider structures that produce exclusion. In this view, disability arises not simply from impairments but from the social organisation and prevailing attitudes that determine whether such impairments are accommodated or neglected (Oliver, 1990, 1996). Disability, in this sense, is constituted by patterns of systemic disadvantage, exclusion, and stigma directed at those with impairments.[2]

A common critique of the social model is that it either underestimates the extent to which impairments can constitute genuine disadvantages in themselves (Terzi, 2004; Anastasiou & Kauffman, 2013), or implies that in a society entirely free of prejudice and exclusion, disability would disappear, leaving only impairments (Barnes, 2016). The latter implication becomes less troubling, however, once disability is understood as a modality of oppression rather than a mere description of bodily difference. In this view, it is coherent to hold that in an oppression-free society impairments might persist as forms of embodied human variation, while no one would be disabled in the sense of being systematically excluded from opportunities for self-development on the basis of those impairments. This does not deny that impairments may involve functional constraints or distinctive challenges. It insists, rather, that such constraints are not themselves oppressive but become so only under specific social, political, and institutional conditions.

In the social model, disability emerges from the interaction between an impairment and a social environment that renders that trait a source of vulnerability to injustice. Whether a characteristic counts as an impairment, and whether it becomes disabling, cannot be determined in abstraction from the social, cultural, and material conditions in which it is embedded. What is experienced as an impairment in one context may not be perceived as such in another, just as the barriers to self-development associated with a given trait vary significantly with prevailing social attitudes, institutional arrangements, and available resources. Disability is therefore not reducible to individual biology but is co-constituted through the interaction between embodied difference and unjust social structures.

Consider Andy, a person with congenital amelia. Andy possesses what is typically classified as an impairment, insofar as this trait diverges from the widely shared anatomical standard of having two arms. On its own, however, this physical difference does not necessarily preclude a full life. Much depends on contextual factors such as available technologies, environmental accessibility, and social attitudes, which can either mitigate or exacerbate any limitations Andy might encounter.

Now imagine Andy situated within a social arrangement not unlike those common in most contemporary societies. In such a context, the absence of arms may become a source of substantial disadvantage – not because of the physical condition itself, but because the world is organised around the assumption of a normative body. Everyday tasks, public spaces, employment systems, and social expectations are designed for people with two upper limbs. The misalignment between Andy's embodiment and the prevailing social infrastructure can therefore result in exclusion, dependence, or diminished opportunities – disability in the socio-political sense.

The point is not to deny the reality of physical difference, but to emphasise that impairment translates into disability only under particular social and material conditions. The challenges Andy faces are not merely medical or personal but structural and political. A society committed to accessibility, accommodation, and inclusion could substantially reduce these barriers, enabling Andy to develop on their own terms. The ethical and political significance of the distinction between impairment and disability lies precisely here – in its dependence on social organisation. As will become clear, this dependence is equally important when analysing forms of oppression that arise wherever embodied vulnerability meets environments, infrastructures, and norms not designed to accommodate it, including in the case of nonhuman animals.

3 Oppressing the Impaired

People with impairments experience oppression in varied and intersecting forms. One familiar mode is linguistic oppression: the everyday vocabulary and metaphors that frame impairments as misfortune, tragedy, or burden. These are not accidental slips of language; they reflect and reproduce entrenched assumptions that lives with impairments are inherently less meaningful, less fulfilling, or less worth living.

Layered onto this is a pervasive cultural judgement that ranks bodies with impairments below non-impaired ones – not only in terms of economic productivity but across nearly every register of human value. These bodies are cast as less beautiful, less capable of intimacy or insight, and ultimately less entitled to flourishing and development.

This evaluative hierarchy does more than stigmatise. It polices who is permitted to develop their capacities and to be visible. It marks certain forms of embodiment as deviations from the norm – lives to be pitied or fixed rather than respected and supported. In this way, the oppression of people with impairments is not merely attitudinal or interpersonal; it is structural, woven into aesthetic, moral, and epistemic frameworks that determine whose forms of life are recognised as fully human.

The structural character of this oppression becomes especially clear when the abstract reasoning of philosophical argument encounters the lived experience of those it purports to assess, as Harriet McBryde Johnson's exchange with Peter Singer illustrates with precision.

In his *Practical Ethics*, Singer defends the permissibility of euthanising infants with severe impairments on the grounds that, given certain assumptions about welfare, suffering, and parental preferences, such lives may be judged to contain less well-being than alternative outcomes, and that parents may therefore permissibly choose to end that infant's life in the expectation of having a future child with better prospects for well-being (Singer, 2011, ch. 7). Singer's argument proceeds by weighing suffering, quality of life, and parental preferences, concluding that some lives to be lived with severe impairments may permissibly be ended because doing so would maximise overall well-being.[3]

Harriet McBryde Johnson, a disability rights lawyer and activist who met Singer in person and engaged him directly on these views, captured the logic of this position with devastating economy: "He insists he doesn't want to kill me. He simply thinks it would have been better, all things considered, to have given my parents the option of killing the baby I once was" (Johnson, 2003). The "*all things considered*" clause plays a crucial epistemic role here. It signals a summative judgement that purports to bring moral reasoning to closure, suggesting that the relevant values and interests have been exhaustively identified, weighted, and balanced. Yet this appearance of completeness conceals a set of normative assumptions about whose experiences count, which goods matter, and who is entitled to make such assessments. The judgement presents itself as neutral while encoding a specific and contestable picture of what makes a life worth living.

In her 2003 *New York Times* essay, "Unspeakable Conversations," Johnson describes what it is like to be the object of this reasoning. During their meeting, Singer presented a hypothetical intended to support his claim that disability is generally correlated with diminished joy: imagine a disabled child on the beach, watching the other children play. Johnson's response was immediate: as a little girl playing on the beach, she had been aware that some people felt sorry for her, that she was not frolicking with the same level of energy as other children. This had annoyed her then and still did. She then described how she had enjoyed playing on the beach without walking, standing, or running (Johnson, 2003).

What Johnson's testimony exposes is not Singer's personal failure of imagination but the ableist framework that makes his hypothetical seem self-evident. The image of the disabled child watching from the margins – passive, excluded, diminished – is not a neutral description of disability but a projection of able-bodied norms onto an embodied experience that does not conform to them. Joy, self-development, and meaningful life can take multiple embodied forms; Singer's hypothetical forecloses this plurality before it can be recognised. Rather than internalising the projected deficit, Johnson names it as the source of alienation, not the impairment itself, but the social gaze that constitutes it as lack.

This exchange should not be read as a personal misunderstanding between two individuals. What it dramatises is the capacity of socially privileged positions to render certain lives legible only through the lens of deficit, tragedy, or burden. These representations are not isolated opinions but expressions of a hierarchical order reproduced across educational, medical, philosophical, and cultural domains – one that embeds the assumption that some lives are more valuable, more worthy of recognition, and more entitled to resources than others. Dominant evaluative frameworks do not merely misrepresent disabled lives; they actively marginalise them by situating people with disabilities at the edges of full moral and political consideration. What Singer presents as an objective evaluation of well-being is, as Johnson's testimony makes clear, a failure of attentiveness, a refusal, enabled by structural privilege, to recognise people with disabilities as epistemic and moral equals.

To grasp how these hierarchies are stabilised and reproduced at a structural level, I turn to one of the most entrenched sources of ableism in contemporary societies: the capitalist mode of production. Although the relationship between disability and capitalism has been explored across diverse intellectual traditions, its conceptual roots can be traced most clearly to Marxian analyses of labour, productivity, and value. Capitalist systems do not merely shape material conditions; they also organise normative standards of bodily capacity, efficiency, and autonomy – standards that systematically marginalise and devalue lives lived with impairments.

One of Marx's central insights was that capitalism is a system organised around the pursuit of wealth through the accumulation of capital. Wealth, in this framework, is primarily generated through productive labour – that is, labour that contributes to surplus value within systems of market exchange. As a result, individuals unable to engage in such labour – whether due to physical, mental, or social constraints – are rendered less valuable in economic terms and, in many contexts, come to be treated as possessing diminished social worth. It is no coincidence that the concept of disability gained public prominence during the Industrial Revolution, precisely when capitalist production intensified and normative standards of bodily productivity became institutionalised (Furuoka, 2024). This was also the historical moment when the modern idea of normality emerged, tied to developments in statistical science and medical classification (Davis, 1997; Pfeiffer, 2001). The structural effects remain palpable today, as people with impairments continue to exhibit systematically lower rates of labour force participation than their non-disabled counterparts (World Health Organization & World Bank, 2011).

Building on this analysis, the critique that matters most here concerns the structural obstacles capitalism places on people with impairments to participate in the labour force. This problem has been influentially analysed by Marta Russell, who, extending the social model into a materialist critique, argues that disability "is a social creation which defines who is offered a job and who is not, and what it means varies with the level of economic activity" (Russell & Malhotra, 2019, p. 19). Within capitalist systems, those whose adaptation to production processes is deemed too costly are excluded from the workforce. People with disabilities, in particular, are cast as burdens – too expensive to be profitable. The disabled body, in this analysis, is produced by capitalism's economic structure in ways that serve the interests of a small capitalist class while generating enormous wealth for it (Russell, 2019, p. 28).

The injustice people with impairments face is not reducible to individual prejudice but emerges from an economic structure oriented toward maximising profit by calibrating production costs against extractable value. Within this logic, impairments become closely associated with unprofitability. This economic rationality underpins the dehumanising comments made by Lewis Schaffer at the outset of this chapter. Such remarks are not merely crude provocations; they crystallise a broader evaluative framework that organises human worth around economic productivity and entrenches hierarchies of exclusion. As Russell observes, it is "neither arbitrary nor irrational that disabled persons have been excluded from education, transportation, and other social spheres. Rather, it is logical that such a state of affairs would exist as long as disabled persons have little value as workers to the capitalist class" (Russell, 2019, p. 36). The consequences are predictable: people with impairments are rendered less employable, more vulnerable to poverty, and systematically pushed to the margins of social and economic life.

State responses to these structural injustices often reinforce, rather than dismantle, the underlying logic of capitalist ableism. In Chile, for instance, it was legal until 2017 to employ "people with disabilities" without adhering to minimum wage regulations. Article 16 of Law 18.600 explicitly stated that in employment contracts entered into by persons with mental disabilities, a remuneration freely agreed upon by the parties could be stipulated, with minimum wage rules inapplicable. While plainly discriminatory and unjust, this provision was defended as a pragmatic incentive to encourage companies to hire workers with impairments. Yet the rationale reveals a troubling logic: people with impairments could be included in the workforce only if their labour was discounted, that is, only insofar as their presence did not undermine profit margins. Reduced productivity was thus offset not through collective responsibility or structural adjustment, but by sanctioning exploitation.

In 2017, this provision was repealed and replaced by Law 21.015, which establishes mandatory hiring quotas for "people with disabilities" in both the public and private sectors and explicitly prohibits discriminatory wage practices. While this framework marks a significant improvement, its orientation remains tied to the imperatives of productivity. Scholars have argued that such policies tend to funnel people with impairments into low-skilled, low-wage labour markets, effectively minimising the social and economic costs of disability rather than transforming the structures that marginalise these lives (Soldatic & Chapman, 2010). In the Chilean context, some analysts contend that these policies introduce state-driven technologies of inclusion that reinforce ableist ideals of independence, autonomy, and productivity (Serrano, 2014; Núñez-Parra, 2020). Reforms of this kind do not eradicate structural oppression; they reconfigure it. People with impairments remain positioned within a hierarchy of social worth, their inclusion contingent on their alignment with capitalist norms of economic efficiency.

The Chilean case is instructive precisely because it represents the reformist limit of the capitalist response to ableism: inclusion is extended, but only on terms that preserve the productive logic underlying exclusion. As the analysis that follows shows, the same logic operates in industrial animal agriculture – but without even the pretence of inclusion, and without the possibility of legal recourse or political contestation.

4 Disabling Other Animals

The capitalist mode of production not only exploits labour; it also determines whose lives are recognised as valuable and whose are rendered disposable. Within this framework, disability is construed as a site of exclusion, insofar as people with impairments are marginalised for failing to meet normative standards of productivity and efficiency. The same evaluative logic governs the treatment of nonhuman animals, whose bodies are commodified, regulated, and discarded according to the imperatives of profit. But in the animal case, this logic operates with a specificity that demands closer attention: the very bodies that capital discards as deficient are bodies that capital itself has produced.

The concept of impairment operates differently when extended to the animal case. In the human case developed above, impairment refers to a bodily or cognitive characteristic that a being possesses prior to and independently of social interpretation. The social model's crucial insight is that this characteristic becomes disabling not through its own nature but through its encounter with social structures, norms, and environments that fail to accommodate it. In the animal case examined here, the relationship between body and social structure is even more direct: the characteristics that render certain animal bodies nonviable within industrial agriculture are not pre-existing traits that the system fails to accommodate. They are traits the system itself has produced or selected for – through selective breeding, genetic manipulation, and reproductive engineering – in order to maximise productive output. The concept of structurally imposed impairment captures this: a condition in which the industrial system both engineers bodies to serve its purposes and simultaneously classifies those same bodies as deficient whenever they fail to perform those purposes completely. The body is shaped by capital and then discarded by it.

In one important respect, this makes the animal case more extreme than the human case. Human disability activists contest the social structures that render impairments disabling; animals in industrial agriculture have no such recourse. But it also makes the analytical connection more direct: ableism names a mode of valuation that ranks bodies according to norms of function, efficiency, and adaptability. As Bob Torres observes in his analysis of racism and speciesism, systems of oppression are "woven into our mental, social, and economic machinery, and reproduced through the interaction of these parts – it is a structural aspect of the political-economic order" (Torres, 2007, p. 9). Ableism operates through the same structural machinery, and the same logic operates across species: under capitalism, bodies that fail to meet productive norms are sorted into the same hierarchy of expendability, whether human or nonhuman. Contesting these norms therefore requires not merely attitudinal change but the dismantling of the spheres of oppression sustained by capitalist modes of production and valuation.

The first and most numerically significant case is the mass culling of male chicks. Each year, an estimated six billion male chicks are killed within hours of hatching, typically by gassing, grinding, or cervical dislocation (Banis, 2018 for scholarly analysis of methods and alternatives, see Krautwald-Junghanns, et al., 2018). Their only deficiency is economic: they cannot lay eggs and do not yield profitable meat. Within this system, they are produced only to be discarded. This practice crystallises a capitalist logic of oppression in which lives are valued solely by their productive contribution. Male chicks are not incidentally killed; their elimination is structurally embedded in the industry's logic of efficiency. In this way, they embody the animal analogue of what Marta Russell (2019) identifies in human contexts as capitalism's production of surplus bodies – lives rendered disposable by the system that created them. Like people with impairments excluded from waged labour because they are deemed unproductive or too costly, male chicks are constructed as surplus. But here the analogy reaches its limit: where people with impairments may be marginalised and excluded, male chicks are simply destroyed.

Viewed through Iris Marion Young's (1990) five faces of oppression, male-chick culling represents both violence – the deliberate destruction of lives rendered unacceptable within economic calculation – and powerlessness, since neither the chicks nor their mothers have any standing to resist or alter their fate. Above all, it exemplifies how hierarchical standards of worth foreclose any possibility of self-development. These animals are denied not only the conditions of development but existence itself, solely because their sex does not conform to the demands of profit.

A second set of practices arises, paradoxically, in response to the ethical and logistical challenges posed by this mass culling. Because routine elimination casts a long moral and public-relations shadow over the industry, new biotechnologies have been developed to avoid post-hatch killing by intervening earlier in reproduction. One such method is *in-ovo* sex determination, which tests embryos around the ninth day of incubation to identify sex-specific biomarkers, allowing male embryos to be discarded before hatching. A more radical technique involves genetic modification to prevent male embryos from developing altogether.

On the surface, these innovations seem to represent moral progress, insofar as they reduce certain forms of visible suffering. Yet their underlying rationale reveals the persistence of structural oppression. These technologies are celebrated not because they affirm the moral value of animals but because they manipulate animal reproduction in ways that better align with human economic interests. Their justification is framed in entirely anthropocentric terms – reducing energy consumption, lowering greenhouse gas emissions, minimising labour costs, and smoothing consumer anxieties. As Wouter Bruins, co-founder of In-Ovo, admitted, many farmers would prefer to stop culling, but their motivations are practical and financial rather than ethical – the economics of the procedure, costing a few pence and saving slightly less, are offered as decisive justifications (Fleming, 2016).

It is here that the concept of anticipatory ableism becomes relevant. Anticipatory ableism describes the logic through which bodies that would later be deemed inefficient, dependent, or unprofitable are eliminated in advance, ensuring that only those conforming to capitalist norms of productivity are ever permitted to exist. It is the temporal extension of the ableist metric of worth: not merely culling nonconforming bodies after the fact, but engineering existence itself so that nonconformity never arises. *In-ovo* sex determination and the pre-hatch elimination of male embryos are paradigmatic instances of this logic. The system does not wait to assess whether a given life can be made productive; it forecloses the question entirely by ensuring that the nonviable body never comes into being. This is structural oppression operating not merely on lives but on the conditions of life itself.

Capitalist agriculture does not renounce the privilege of controlling animal lives through these techniques; it refines its tools of oppression. Male embryos remain disposable. Their destruction is displaced to an earlier stage of reproduction and rebranded as more efficient, but the structure of oppression remains intact. Animals are treated not as beings with their own capacities for self-development but as material to be optimised, managed, or eliminated depending on their utility. Far from transcending the ableist-capitalist metric of value, these biotechnologies entrench it – functioning as more sophisticated means of erasing lives that fail to serve the imperatives of profit.

The lives of hens make this dynamic still more visible. Their bodies are not merely used but relentlessly extracted: many lose the ability to walk or stretch within the confines of their cages; their eyesight deteriorates under the corrosive effects of ammonia from accumulated waste; and their beaks are systematically mutilated to suppress the injuries that would otherwise result from the stress of confinement. What is celebrated as efficient production is rather the rationalisation of misery, an effort to optimise the endurance of oppressed bodies while concealing the structural violence at its core.

A further illustration of this dynamic can be found in the selective breeding of featherless chickens, developed at the Hebrew University of Jerusalem through crossbreeding using a naturally occurring scaleless mutation and widely publicised in the early 2000s (Cahaner et al., 2008; Traubman, 2002). Marketed as an environmentally friendly and cost-saving innovation, the featherless chicken was designed to reduce the expenses of plucking in slaughterhouses, improve feed conversion rates, and facilitate faster growth in hot climates by eliminating the energy birds expend on thermoregulation. The naked body of the animal was deliberately produced to optimise industrial efficiency.

From a capitalist standpoint, this intervention was framed as progress: a body redesigned to meet the demands of production, stripped of features deemed inconvenient to the machinery of profit. Viewed through the lens of oppression, however, the featherless chicken exemplifies the extent to which anthropocentric privilege manipulates and reconfigures animal lives. The bird's most basic biological features – including the protective covering that enables thermoregulation, resistance to parasites, and other vital functions – are recast as inefficiencies to be eliminated. The result is an animal bred into heightened vulnerability: exposed to sunburn, insect infestation, and climatic extremes, and rendered dependent for its survival on the tightly controlled environments of industrial farming. The title of this chapter – disabled by design – is perhaps most literally illustrated here. The featherless chicken has not encountered an unaccommodating social environment; it has been designed into dependency by a productive system that simultaneously requires its existence and forecloses its flourishing.

What this case makes visible is the intimate entanglement of ableism and speciesism under capitalist logics – what Taylor (2017) has identified as the shared structures of oppression linking disability and animal exploitation. Like people with impairments rendered unfit for the labour market, featherless chickens are designed into dependency, their lives calibrated around the requirements of productivity and control rather than their own flourishing. In both cases, the hierarchy of value is dictated by anthropocentric privilege: human interests set the standards of normalcy, and any body that deviates is either eradicated, reconfigured, or permanently subordinated. The featherless chicken is the crystallisation of a broader structure – a sphere of oppression in which technological manipulation masquerades as innovation, and in which lives are systematically redesigned to better serve the priorities of capital. The violence here is not only physical but epistemic: it erases the possibility of recognising animals as beings with their own capacities for self-development, rewriting their very biology to mirror back the logic of human supremacy.

5 Conclusion

This chapter has argued that ableism is not an incidental prejudice or marginal cultural attitude but a constitutive logic of capitalist societies, one that organises value, productivity, and disposability not only within the human domain but across species lines. By equating worth with efficiency, independence, and economic contribution, ableism legitimises the exclusion of people with disabilities from full participation in social and economic life. The same logic, operating through the same institutional and evaluative machinery, governs the treatment of nonhuman animals in industrial agriculture, where bodies are manipulated, redesigned, and eliminated according to their productive utility. What connects the Schaffer remark at the opening of this chapter to the mass culling of male chicks and the breeding of featherless chickens is not coincidence but structure: a shared evaluative grammar that renders certain lives disposable by measuring them against norms of productivity they were never designed to meet.

The practices examined here – the culling of male chicks, *in-ovo* sex determination, and the selective breeding of featherless chickens – make visible how oppression operates through multiple overlapping faces. Following Young's framework, we find violence in the literal destruction of lives rendered economically unacceptable; powerlessness in the total absence of any agency to resist industrial imperatives; exploitation in the continuous extraction and optimisation of animal bodies as machines of production; and marginalisation in the systemic exclusion of animals' needs, capacities, and forms of life from moral and political consideration. Crucially, the biotechnological innovations examined here do not represent a departure from this structure. They refine it. By displacing culling to an earlier stage of reproduction, or by breeding bodies that conform more perfectly to productive norms, they extend and intensify the ableist metric of worth while presenting themselves as solutions to it. What emerges is not moral progress but the modernisation of misery – oppression rendered more efficient, less visible, and therefore more difficult to contest.

The concept of anticipatory ableism captures how far this logic extends. It names the temporal dimension of capitalist oppression: not merely eliminating bodies that fail to meet productive norms after they have come into being, but engineering existence itself so that only conforming bodies are ever permitted to arise. This is structural oppression at its most fundamental level, operating not on lives already lived but on the conditions under which life is possible. *In-ovo* sex determination and the pre-hatch elimination of male embryos are its paradigmatic instances in the animal domain; the long history of eugenic reasoning about disability in the human domain shows that the logic is not confined to one side of the species boundary.

The chapter's broader conclusion is that ableism and speciesism are not merely analogous but structurally co-constituted within capitalist frameworks. This claim converges with Taylor's (2017) analysis of disability and animal liberation as entangled forms of subjugation, and develops it in a specific political economy register. This is not to claim that capitalism originates these hierarchies – both have deeper historical and metaphysical roots – but that it provides the shared institutional logic through which they are intensified, aligned, and rendered mutually reinforcing. Within this logic, both human and nonhuman lives are subjected to the same evaluative grammar: those who cannot be rendered productive are classified as surplus, and those classified as surplus are rendered disposable. Confronting ableism is therefore not only a matter of justice for persons with disabilities. It is a way of challenging the broader political economy that naturalises the disposability of vulnerable lives and of recognising that the struggle against that economy cannot be won on one side of the species boundary alone.

Notes

1 For this see, for example, Amundson (2000) on the biological basis of functional normality, and Barnes (2016, ch. 1) on under- and over-inclusiveness.

2 This formulation traces back to the Union of the Physically Impaired Against Segregation (UPIAS), whose *Fundamental Principles of Disability* (1976) laid the groundwork for what later became known as the social model. UPIAS famously defined disability as a form of social oppression imposed upon people with impairments through exclusionary social arrangements: "Disability is something imposed on top of our impairments by the way we are unnecessarily isolated and excluded from full participation in society … Physical disability is therefore a particular form of social oppression." While subsequent disability theorists have criticised early versions of the social model for underplaying embodiment and lived experience, its core insight – that disability is produced through social organisation rather than bodily difference alone – remains foundational for structural analyses of oppression.

3 "Parents may, with good reason, regret that a disabled child was ever born. In those circumstances, the effect that the death of the child will have on its parents can be a reason for, rather than against, killing it" (Singer 2011, p. 161).

References

Amundson, R. (2000). Against normal function. *Studies in History and Philosophy of Biological and Biomedical Sciences*, 31 (1), 33–53.

Anastasiou, D., & Kauffman, J. M. (2013). The social model of disability: Dichotomy between impairment and disability. *Journal of Medicine and Philosophy*, 38(4), 441–459.

Banis, D. et al. (2018). Technology may save six billion male chicks a year from the grinder. *Kinder World*. https://kinder.world/articles/solutions/technology-may-save-six-billion-male-chicks-a-year-from-the-grinder-19477.

Barnes, E. (2016). *The minority body: A theory of disability*. Oxford University Press.

Bolton, D. (2008). *What is mental disorder? An essay in philosophy, science, and values*. Oxford University Press.

Boorse, C. (1975). On the distinction between disease and illness. *Philosophy & Public Affairs*, 5(1), 49–68.

Boorse, C. (1977). Health as a theoretical concept. *Philosophy of Science*, 44(4), 542–573.

Boorse, C. (1997). A rebuttal on health. In J. M. Humber & R. F. Almeder (Eds.), *What is disease?* (pp. 1–134). Humana Press.

Campbell, F. K. (2009). *Contours of ableism: The production of disability and abledness*. Palgrave Macmillan.

Cahaner, A. et al. (2008). Effects of the genetically reduced feather coverage in naked neck and featherless broilers on their performance under hot conditions. *Poultry Science*, 87(12), 2517–2527. doi:10.3382/ps.2008-00284.

Davis, L. J. (1997). Constructing normalcy. In L. J. Davis (Ed.), *The disability studies reader* (pp. 9–28). Routledge.

Fleming, N. (2016, May 27). Chicken embryo tests can prevent practice of gassing billions of cockerels. *The Guardian*. www.theguardian.com/environment/2016/may/27/chicken-embryo-tests-can-prevent-practice-of-gassing-billions-of-cockerels.

Furuoka, F. (2024). *Disability and employment: Towards a humanistic economy*. Springer.

Garland-Thomson, R. (2002). Integrating disability, transforming feminist theory. *NWSA Journal*, 14(3), 1–32.

Jenkins, K., & Webster, A. (2020). Disability, impairment, and marginalised functioning. *Australasian Journal of Philosophy*, 99(4), 730–747.

Johnson, H. M. (2003, February 16). Unspeakable conversations. *The New York Times Magazine*. www.nytimes.com/2003/02/16/magazine/unspeakable-conversations.html.

Krautwald-Junghanns et al. (2018). Current approaches to avoid the culling of day-old male chicks in the layer industry, with special reference to spectroscopic methods. *Poultry Science*, 97(3), 749–757. doi:10.3382/ps/pex389.

Núñez-Parra, L. (2020). Discapacidad y trabajo. *Revista Nómadas*, 52, 61–79.

Oliver, M. (1990). *The politics of disablement*. Macmillan.

Oliver, M. (1996). *Understanding disability: From theory to practice*. Macmillan.

Pfeiffer, D. (2001). The conceptualization of disability. In S. N. Barnatt & B. M. Altman (Eds.), *Exploring theories and expanding methodologies: Where we are and where we need to go* (pp. 29–52). JAI Press.

Pring, J. (2025, July 3). GB News says it has nothing to apologise for. *Disability News Service*. www.disabilitynewsservice.com/gb-news-says-it-has-nothing-to-apologise-for-after-guest-suggests-starving-disabled-benefit-claimants/.

Russell, M. (2019). *Capitalism and disability: Selected writings by Marta Russell* (D. Goodley, Ed.). Haymarket Books.

Russell, M., & Malhotra, R. (2019). Capitalism and disability. In M. Russell, *Capitalism and disability: Selected writings by Marta Russell* (pp. 15–30). Haymarket Books.

Serrano, P. (2014). *Producción de subjetividad. Revista Chilena de Terapia Ocupacional*, 14 (2), 119–136. doi:10.5354/0719-5346.2014.35715.
Singer, P. (2011). *Practical ethics* (3rd ed.). Cambridge University Press.
Soldatic, K., & Chapman, A. (2010). Surviving the Assault? The Australian Disability Movement and the Neoliberal Workfare State. *Social Movement Studies*, 9(2), 139–154. doi:10.1080/14742831003603299.
Taylor, S. (2017). *Beasts of burden: Animal and disability liberation*. The New Press.
Terzi, L. (2004). The social model of disability: A philosophical critique. *Journal of Applied Philosophy*, 21(2), 141–157.
Torres, B. (2007). *Making a killing: The political economy of animal rights*. AK Press.
Traubman, T. (2002, May 23). The naked truth about the naked chicken. *Haaretz*. www.haaretz.com/israel-news/culture/2002-05-23/ty-article/the-naked-truth-about-the-naked-chicken/0000017f-db2c-d856-a37f-ffec19ce0000.
Union of the Physically Impaired Against Segregation. (1976). *Fundamental principles of disability*. UPIAS.
World Health Organization & World Bank. (2011). *World report on disability. World Health Organization*. www.who.int/publications/i/item/9789241564182.
Young, I. M. (1990). *Justice and the politics of difference*. Princeton University Press.
Young, I. M. (2011). *Responsibility for justice*. Oxford University Press.

4

OPPRESSIVE MIGRATION, BORDERS, AND SLAUGHTERHOUSES

1 Introduction

The logics of valuation and disposability analysed in the previous chapter do not end with disability or industrial animal agriculture. They reappear, with striking force, in the governance of migration. Contemporary border regimes operate through classificatory practices that sort lives according to productivity, adaptability, and perceived burden, rendering some individuals admissible while exposing others to abandonment, containment, or death. These regimes do not merely regulate movement; they actively organise vulnerability by determining where, how, and under what conditions harm may occur. In doing so, they entangle the fates of human migrants and nonhuman animals, whose lives are similarly subjected to administrative neglect, infrastructural violence, and forms of managed exposure. Migration thus emerges as a further site in which capitalism, ableism, and speciesism converge, extending the same evaluative grammar that normalises exclusion and disposability across species lines.

When vulnerability is organised in this way, death does not appear as an accident but as a predictable outcome of governance. Since 2014, more than 56,000 people have died along global migration routes (Aldridge et al., 2018; Poole et al., 2020; Kerwin & Martínez, 2024; Passarelli-Araújo, 2025) not as the result of natural hazards or unfortunate miscalculations, but as the foreseeable consequence of policies that externalise borders – extending immigration controls into transit countries and international waters – fragment responsibility, and render some lives politically expendable. The fact that most of these deaths remain unidentified is not accidental: anonymity is one of the mechanisms through which migration governance transforms persons into administrable losses, absorbed into what functions as a global infrastructure of disposability rather than a series of isolated humanitarian failures.

DOI: 10.4324/9781003434856-5

This chapter is not concerned with migration as a general or transhistorical phenomenon – humans have been migrating for approximately 70,000 years. It focuses instead on migration as a site of oppression, in which unjust social, economic, and political conditions compel movement and then systematically expose those who move to heightened risks of death, degradation, and erasure. Such migration is not incidental or exceptional but characteristic of contemporary global arrangements that rely on mobility while simultaneously criminalising and endangering it.

The preceding chapter on disability showed how modern institutions manage vulnerability through regimes of neglect and conditional inclusion. Migration reveals the same governing logic operating through a different spatial strategy. Where disability governance tends to confine vulnerable bodies within institutional settings marked by abandonment, migration externalises vulnerability, forcing it into deserts, seas, and border zones where harm can be disowned while remaining structurally produced. In both cases, injustice is defined not by overt exclusion but by the systematic organisation of exposure to harm under conditions of diminished political accountability.

These oppressive logics do not stop at the boundaries of the human. Approached through a non-anthropocentric lens, migration governance appears as part of a broader set of practices that regulate movement and vulnerability across species lines. Border regimes, labour markets, and systems of food production generate interlocking sites of oppression in which the lives of migrant humans and nonhuman animals are jointly shaped by logics of control, exploitation, and disposability. This chapter argues that these connections are not merely analogical but structural: migration regimes and animal exploitation are co-constituted within political and economic systems that depend on the managed vulnerability of both.

The argument unfolds in three steps. First, the chapter clarifies which forms of migration are at stake, distinguishing structurally compelled mobility from more general accounts of movement and introducing the concept of oppressed migrants. Second, it examines the US–Mexico border as a paradigmatic site in which human and animal movement is simultaneously controlled, policed, and punished. Third, it turns to the slaughterhouse and industrial dairy farm, where migrant labour is indispensable to animal killing and exploitation, rendering these sites emblematic of how capitalist food production entwines human and animal suffering within the same regime of disposability. Taken together, these cases show how migration operates as a multispecies domain of injustice – one that demands a rethinking of borders, labour, and responsibility beyond the limits of anthropocentric political theory.

2 Migration and Structural Oppression

Under international law, there is no universally accepted definition of the term migrant. The International Organization for Migration (IOM) offers an intentionally expansive account, defining migrants as individuals who move away from their usual place of residence, whether within a state or across international borders (Sironi et al., 2019, p. 132). This umbrella category encompasses migrant workers, smuggled migrants, refugees, international students, and others. Although often presented as descriptively neutral, such definitional breadth is politically and normatively consequential. By flattening radically different forms of mobility into a single category, broad definitions obscure the conditions under which movement is compelled, dangerous, or systematically harmful, thereby enabling institutions to evade responsibility for the injustices that structure much contemporary migration.

To analyse migration as a site of oppression rather than as a mere pattern of movement, a more discriminating framework is required. This chapter proposes a model that understands migration as unfolding across four analytically distinct stages, each constituting a sphere of oppression in its own right. This staged approach is normatively indispensable because it shows that injustice is not confined to a single moment – departure, transit, or arrival – but is reproduced and transformed across the entire migratory trajectory. It also reveals how the mechanisms that constrain and endanger migrant humans frequently intersect with practices that entangle nonhuman animals, a connection largely absent from dominant migration theories.

The first stage consists of the oppressive conditions that compel departure from countries or regions of origin: war, political violence, systemic discrimination, economic exploitation, and climate-related degradation, often rooted in colonial legacies and entrenched global inequalities. In such contexts, remaining in place becomes incompatible with a minimally decent life. The second stage centres on the journey itself, during which migrants are exposed to distinctive forms of violence, deprivation, and death. Deserts, seas, and border zones function as spaces where vulnerability is deliberately intensified – as illustrated by Carolina's experience in Chapter 2. The third stage emerges upon arrival, when migrants typically lack enforceable protections and are exposed to heightened risks of abuse by border authorities, law enforcement agencies, employers, and private actors.[1] The fourth and final stage concerns those who survive and achieve relative stability but remain subject to durable forms of exclusion, precarity, and marginalisation, often persisting across generations.

Across these stages, oppression operates less through outright exclusion than through the systematic organisation of risk, insecurity, and diminished agency. Migration governance actively produces vulnerability by distributing exposure to harm in ways that are politically manageable and morally normalised. Recognising this continuity is essential for understanding migration as a structural injustice rather than a sequence of unfortunate episodes.

The IOM's broad definition, reconsidered in light of this framework, proves inadequate for normative analysis. By encompassing forms of mobility that are voluntary, privileged, or professionally rewarded, it obscures the moral distinction between movement undertaken under conditions of freedom and movement compelled by injustice. A narrower alternative is offered by the 1951 Refugee Convention, which defines refugees as persons fleeing persecution on the grounds of race, religion, nationality, membership of a particular social group, or political opinion. Unlike the IOM's definition, this account explicitly acknowledges that some forms of migration are forced rather than chosen, and thus tracks injustice more closely. Yet the Convention remains deeply restrictive. By anchoring protection to persecution rather than to broader forms of structural oppression, it excludes many of the most pervasive and foreseeable drivers of contemporary displacement.

This limitation is especially evident in cases of migration induced by anthropogenic environmental collapse. Droughts, wildfires, flooding, soil degradation, and land subsidence – phenomena directly linked to human activity – are increasingly rendering entire regions uninhabitable. Individuals displaced under such conditions often fall outside existing legal protections, despite the fact that their capacity for self-development is systematically constrained. As argued in Chapter 2, such constraints are paradigmatic of oppression regardless of whether those affected articulate their experience in explicitly political terms. The failure to recognise environmentally displaced persons as subjects of injustice is therefore not a mere classificatory oversight but a moral failure to acknowledge the unjust conditions that make continued residence untenable.

These limitations indicate the need for a different analytical category: oppressed migrants. By this term, the chapter refers to individuals compelled to leave their place of residence – whether within or across national borders – because of structural forms of oppression, including persecution, economic exploitation, systemic discrimination, political violence, and human-induced environmental collapse. This category does not coincide neatly with existing legal classifications; its purpose is not to replace them but to provide a normative lens capable of identifying when migration itself expresses injustice.

Crucially, the status of an oppressed migrant is not determined at the moment of departure. Migration exposes individuals to shifting and overlapping spheres of oppression over time. A person may initially migrate in search of improved economic opportunities under conditions that appear voluntary, only to encounter severe exploitation, coercion, or violence upon arrival – circumstances that render them oppressed migrants even if their initial movement was not directly compelled.[2] Conversely, an individual who flees persecution may later find more supportive conditions abroad. Yet the injustice that necessitated their displacement remains normatively salient: improved circumstances do not erase the wrongs that made departure unavoidable, nor the broader structures that continue to shape migrants' vulnerability.

Understanding migration through shifting exposure to oppression rather than fixed legal status will be central to the arguments that follow. It is also this dynamic, stage-based conception of vulnerability that enables the analysis of how migration regimes entangle human and nonhuman lives within shared structures of control, exploitation, and disposability.

3 Borders as Sites of Animal Oppression

Once the contours of oppressive migration have been established, we can turn to how the experiences of oppressed human migrants are structurally bound up with those of nonhuman animals. This section grounds in concrete practices the concept of interspecies oppression developed in Chapter 2, understood there as a form of structural injustice that suppresses the conditions for self-development through exploitation, marginalisation, powerlessness, cultural imperialism, and violence. Migration regimes provide a particularly revealing lens for this analysis, as they organise vulnerability across species lines while simultaneously obscuring its structural character.

Oppression does not operate in abstraction. It materialises within specific spheres shaped by the convergence of attitudes, institutional arrangements, and coercive practices that sustain hierarchies of privilege. Borders crystallise these dynamics with particular clarity. They are spaces where human vulnerability is actively produced through exclusionary policies, surveillance, and violent enforcement, while simultaneously functioning as ecological barriers that fragment habitats and obstruct the movement of nonhuman animals.

In the mid-2010s, Slovenia erected fences along its border with Croatia to block the movement of migrants and asylum seekers. These fences simultaneously severed a key wildlife corridor, threatening populations of lynx, wolves, and bears. Empirical research shows that such barriers obstruct access to essential resources, reduce effective population sizes, and increase mortality through injury, exhaustion, and habitat fragmentation (Linnell et al., 2016). Animals trapped in these structures often die slowly, revealing how migration control policies impose lethal consequences far beyond their intended human targets.

Comparable dynamics are evident along the Chinese–Mongolian border, where extensive fencing blocks the migratory routes of the Asiatic wild ass, forcing animals to follow fence lines that expose them to heightened poaching and ecological stress. In India, fencing along the Bangladesh border fragments the habitat of wild Asian elephants, disrupting seasonal migrations and intensifying human–elephant conflict (Government of India 2021; Shaffer et al., 2019). In the occupied Palestinian West Bank, Israel's separation wall obstructs the movement of multiple species, including foxes, wolves, and the endangered mountain gazelle. Spanning more than 700 kilometres and reaching heights of over eight metres, the wall contributes to habitat fragmentation, population isolation, genetic bottlenecks, and demographic instability, including Allee effects that reduce survival and reproduction rates in already declining populations (Yom-Tov et al., 2021; IUCN, 2020).[3]

In each case, infrastructures designed to police human movement simultaneously produce new forms of precarity for nonhuman animals. Border regimes do not merely regulate human mobility; they reorganise ecological space in ways that entrench vulnerability across species. These dynamics illuminate a structural pattern that the following section examines in depth through the paradigmatic case of the US–Mexico border.

4 The Case of the US–Mexico Border

The US–Mexico border provides a paradigmatic site for examining how migration regimes organise vulnerability across species lines. As a densely governed space of surveillance and infrastructural violence, the border regulates human mobility while simultaneously shaping the conditions under which both human and animal lives are exposed to harm and death.

Migration across this border is characterised by a profound asymmetry of mobility. US citizens travelling south of the Río Grande typically do so under conditions of ease and legal protection, pursuing leisure, consumption, or cultural experience. Their movement is facilitated and normalised by border institutions. By contrast, more than two million people from Central and South America attempt to travel north each year under conditions of necessity rather than choice, driven by poverty, violence, political instability, and environmental degradation that foreclose the possibility of self-development. Their movement is criminalised, intensively surveilled, and channelled into increasingly dangerous routes. Northbound migration across the US–Mexico border has been aptly described as "egregiously dangerous, inhumane, and horrid" (Reed-Sandoval, 2019, p. 767).

In this context, migrants such as Carolina – the Guatemalan girl whose story opened Chapter 2 – are not positioned as travellers but as disposable and manipulable lives. The border functions as a sphere of oppression that distributes mobility as privilege to some while systematically exposing others to profound vulnerability.

These outcomes are not accidental. Since the 1990s, US policies of prevention through deterrence have deliberately redirected migration flows into remote desert regions, transforming geography itself into an instrument of enforcement. The predictable result has been a dramatic rise in deaths from dehydration, exposure, and drowning. In 2021 alone, at least 650 people died attempting to cross the US–Mexico border – the highest number ever recorded by the International Organization for Migration. Vulnerability here is not a policy failure but a foreseeable outcome of institutional design. Far from a neutral security device, the border wall encodes privilege, exclusion, and disposability into the landscape, enforcing practices that actively produce environments of violence.

4.1 Borders, Ecological Fragmentation, and Animal Oppression

The infrastructures that endanger human migrants also restructure ecosystems, generating parallel and intersecting forms of vulnerability for nonhuman animals. Extending nearly 1,993 miles from the Pacific Ocean to southeast Texas, the US–Mexico border has been transformed into a militarised landscape in which systematic human rights abuses against migrants coexist with large-scale ecological disruption. Approximately 700 miles of this border are segmented by walls and fences of varying forms – from low steel bollards and wire mesh to eight-metre-high steel barriers. Despite their material differences, these structures share a common function: to control and exclude human movement. In doing so, they fragment ecosystems, destroy habitats, and obstruct animal migrations essential for survival.

The ecological consequences are severe. The border wall cuts through some of the most biodiverse regions in North America, threatening species already vulnerable at the edges of their geographic ranges. Jaguars, ocelots, Sonoran pronghorns, and the cactus ferruginous pygmy owl face heightened risks of starvation, genetic isolation, and local extinction as habitats are divided and migratory corridors severed (Greenwald et al., 2017; Peters et al., 2018). More than 700 migratory species of birds, mammals, and insects rely on these borderlands annually, many of which are now forced to navigate landscapes deliberately rendered impassable. Scientific studies consistently show that habitat fragmentation reduces effective population sizes, limits genetic variability, and significantly increases extinction risk (De Angelo et al., 2011; Roques et al., 2016; Harrity et al., 2024). While conservation areas may offer partial protection, habitat connectivity remains the single most important condition for sustaining viable jaguar populations in the region (McCain & Childs, 2008).

The implications are particularly stark under conditions of accelerating climate change. As rising temperatures, prolonged droughts, and shifting ecosystems reshape the Sonoran and Chihuahuan deserts, both humans and nonhuman animals are increasingly compelled to move to survive. For migrants, climate stress compounds existing drivers of displacement, further foreclosing the possibility of remaining in place. For wildlife, altered precipitation patterns and habitat degradation intensify the need for mobility across increasingly fragmented landscapes. Projections suggest that by 2070 a substantial proportion of mammal and bird species will find much of their climatically suitable habitat located beyond their current national ranges (Titley et al., 2021). In this context, border infrastructures collide directly with ecological necessity, transforming climate-induced movement into heightened exposure to harm. What emerges is not a temporary crisis but a durable configuration in which climate change amplifies the border's function as a site of interspecies oppression. The border wall is not merely a political artefact but an ecological instrument of exclusion, binding human and nonhuman lives within the same regime of structurally produced vulnerability.

These policies constitute instances of oppression in the precise sense developed in Chapter 2. Border infrastructures generate this injustice by systematically targeting vulnerable human populations while actively undermining the conditions necessary for nonhuman survival. They instantiate a biased political order that privileges certain lives and forms of mobility while rendering others disposable. As Keith Miyake argues, border control operates through a carceral logic that "partitions and segregates, marking people as inside or outside, and facilitating accumulation through regional differentiation in policies regarding production, labor, and the environment" (Miyake, 2023, pp. 1297–1298). Citizenship and settler-colonial relations are stabilised by elevating some groups while exposing others to abandonment, containment, and death.

This carceral logic extends beyond human communities. It also structures interspecies relations, producing animals as both ecological casualties and destabilised cultural symbols.

The jaguar (*Panthera onca*) offers a particularly vivid illustration. Across the Americas, jaguars have long held deep symbolic significance, appearing as deities, warriors, shamans, and embodiments of power in Indigenous cosmologies. Their presence has shaped human understandings of environment, identity, and coexistence across diverse cultures. Yet the fragmentation of jaguar habitat by the border wall undermines not only biological survival but also the cultural meanings historically woven around the species. The wall erodes a multispecies cultural fabric, displacing nonhuman animals alongside Indigenous peoples and subordinated human migrants.

This cultural erasure is not incidental to the economic logics of capitalism – it is their precondition. Once stripped of symbolic depth and relational meaning, the jaguar becomes available for a different kind of valuation entirely: that of the luxury commodity market. Online retailers openly advertise jaguar cubs for sale, marketing them as luxury companions with assurances of veterinary care, optimal conditions, and an explicit "Easy Return Policy" should the animal fail to meet consumer expectations (Luxury Pet Source n.d.). In this consumerist grammar, the jaguar's symbolic richness is stripped away and replaced by the language of commodities. What once embodied power, relationality, and ecological belonging is reconstituted as a luxury possession for the wealthy. The jaguar's reduction to a commodity is not a separate phenomenon from its ecological and cultural displacement; it is the same logic carried to its conclusion: a being whose world has been destroyed is more easily captured, marketed, and discarded.

Viewed through Iris Marion Young's account of structural injustice, these dynamics generate multiple overlapping faces of oppression. Jaguars are subjected to cultural imperialism, as dominant profit-driven narratives overwrite their historical and relational meanings; to marginalisation, as their lives are rendered expendable within both ecological and economic systems; and to powerlessness, as they are deprived of any agency and reduced to passive recipients of human-imposed practices. Most fundamentally, these processes constitute a direct assault on jaguars as individual subjects, systematically constraining their capacity for self-development – to move, hunt, reproduce, play – in the service of human political and economic interests.

The convergence of militarised borders, ecological fragmentation, cultural erasure, and capitalist commodification produces interlocking spheres of oppression in which both human and nonhuman lives are denied the conditions for self-development. The border is not an exceptional space but a paradigmatic one: a carceral infrastructure that organises mobility, labour, and vulnerability in ways that render certain lives disposable while sustaining the conditions under which others flourish. Yet the border is not the endpoint of this logic. It is, in an important sense, its point of production – the site at which legally precarious, socially invisible, and economically coerced human subjects are generated and then distributed into the labour markets that industrial capitalism requires. Nowhere is this more visible than in the slaughterhouse and the industrial dairy farm.

5 Migration and Oppression in the Slaughterhouse

The border and the slaughterhouse are not separate domains connected by analogy; they are consecutive moments within the same political economy. One generates cheap and vulnerable human labour at the frontier; the other deploys it in the service of animal killing at the interior. Both are governed by the same logic of managed vulnerability, and in both, human and nonhuman lives are subjected to regimes of exposure, containment, and disposability.

Industrial animal agriculture has, over the past century, restructured the biosphere around the large-scale confinement, exploitation, and conversion of animals into commodities – an estimated 80 billion land animals killed annually, with livestock now accounting for approximately 60 per cent of total mammalian biomass on Earth (Bar-On et al., 2018). Sustaining this system at its current scale requires an enormous human workforce, and it is here that the border and the slaughterhouse converge: the same political economy that produces migrant vulnerability at the frontier channels that vulnerability into the labour regime of industrial killing.

Producing this volume of animal flesh is work that most people will not willingly do. Slaughterhouses and meat-packing plants face persistent staffing challenges not because the work is scarce but because it is dangerous, degrading, and deeply aversive. Timothy Pachirat argues that these facilities function as "a place that is no-place" rendering them socially and morally invisible so that the violence required for mass consumption can be delegated to others (Pachirat, 2011, p. 3). First-hand testimonies make this concrete. Workers describe abattoirs as "filthy, dirty places" saturated with animal faeces, blood, and the smell of dying animals from the moment of entry; they recount the routine physical danger of handling large animals in distress, the psychological weight of repetitive killing, and the social stigma that follows them outside the workplace (BBC, 2019).[4] These are not aberrations but characteristic features of labour organised around the industrialised killing of animals, and they signal, already, the emergence of a distinctive sphere of oppression in which human and animal suffering converge.

It is precisely because slaughterhouse work is so widely shunned that migrant workers are systematically drawn into it. With limited legal protections, constrained alternatives, and vulnerability produced at and beyond the border, migrants become the workforce that sustains industrial killing. Their entry into this labour is not best understood as a free occupational choice but as the predictable outcome of structural injustice – and the pattern is global.

Comparative analyses across Germany, the Netherlands, and the United States confirm that immigrant workers dominate the meat industry workforce in every national context studied, facing intersecting precarity rooted in both immigration status and employment conditions (Gottlieb & Holmes, 2022; Gottlieb et al., 2025. See also Fremstad et al., 2020). In the United States, refugee workers have been actively recruited to replace undocumented Latinx workers in meat-packing plants, positioning refugees as essential yet acutely vulnerable labour – their precarity managed and reproduced by an industry that benefits directly from their displacement (Frydenlund & Dunn, 2022). Research on Canadian meat-packing plants extends this finding: even permanent residents formerly resettled as refugees encounter dangerous working environments and limited capacity to exercise labour rights, demonstrating that precarity is not a transitional feature of recent arrival but a structural condition that persists long after legal status is secured (Bragg & Hyndman, 2024). In Germany, the meat industry's economic boom relies heavily on migrant labour, including refugees and recent EU arrivals, who experience systematic exploitation – though research also documents workers' capacity for resistance and collective organising, a reminder that structural subordination does not extinguish agency (Birke & Bluhm, 2020).

The institutional mechanisms through which this vulnerability is produced and maintained are well documented. A 2020 report by the European Federation of Food, Agriculture and Tourism Trade Unions shows that workers in the meat sector are frequently recruited through exploitative intermediaries and opaque subcontracting arrangements, required to pay recruitment fees, and subjected to systematic deductions for accommodation, transport, and work equipment. Housing is often overcrowded, contractually tied to employment, and owned by employers, reinforcing dependence and limiting workers' capacity to exit abusive conditions. Legal loopholes – letterbox companies, bogus cooperatives – are routinely used to suppress wages, evade taxation, and externalise risks onto workers (EFFAT, 2020). In the United Kingdom, employers benefit deliberately from migrants' hyper-flexible work patterns to maintain a cheap food supply, producing a workforce that is structurally invisible precisely because its exploitation is so thoroughly normalised (Lever & Milbourne, 2016).

Migrants thus reach what is often framed as the stage of integration only to find themselves absorbed into an economic structure that converts legal precarity into labour discipline. In Iris Marion Young's terms (Young, 1990), this arrangement embodies not only exploitation but also powerlessness: migrant workers are confined to subordinated positions, deprived of control over the conditions and pace of their labour, and largely excluded from effective political or institutional redress. The dirty, dangerous, and demeaning work of transforming animals into commodities is systematically allocated to those whose vulnerability renders them most exploitable – and least able to refuse.

Within these spaces, the oppression of workers and the oppression of animals are not parallel but entangled. Animals are confined, killed, and dismembered as units of production; workers are marginalised, disciplined, and subjected to hazardous and degrading labour regimes. Both are governed by institutional arrangements that normalise harm, foreclose meaningful alternatives, and render suffering invisible to those who benefit from it. The slaughterhouse is therefore not merely a workplace but a crucible of interspecies injustice – a site where the oppression of human workers and the oppression of animals are sustained by, and mutually reinforce, the same institutional logic.

The dairy industry extends and deepens this logic. Like slaughterhouses, industrial dairies rely on a labour regime structured by precarity: in the United States, immigrant workers constitute approximately 51 per cent of all dairy labour (Adcock et al., 2015). And like slaughterhouses, they are sites where animal bodies are subjected to overlapping forms of oppression that map directly onto Young's five faces. To sustain continuous milk production, cows are forcibly impregnated on an annual basis, and their calves are typically removed within hours of birth – a routine separation that constitutes violence in Young's sense: systemic harm inflicted as part of normal social practice, institutionally legitimised and widely tolerated. Male calves are commonly killed within days as surplus to production requirements or raised for beef or veal under conditions of confinement; female calves are incorporated into the same reproductive regime as their mothers, subjected to repeated cycles of insemination, separation, and milk extraction until declining productivity renders them economically spent and they are sent to slaughter, typically at around six years of age, a fraction of their natural lifespan.

This cycle embodies exploitation precisely: cows' reproductive capacities are appropriated and intensified far beyond species-typical limits, producing milk not for their calves but for human consumption and corporate profit. It also exemplifies marginalisation – cows are systematically denied the opportunity to nurture offspring, form stable social bonds, or engage in behaviours central to their flourishing – and manifests powerlessness, since their reproductive cycles, bodily integrity, and productive lives are wholly determined by external economic imperatives, leaving them no meaningful capacity for refusal.[5] Routine industry practices underscore this condition. Disbudding – the burning or cutting of horn buds in young calves – is commonly defended as a safety measure, yet it inflicts severe pain and long-term distress. As Neave et al. (2013) demonstrate, disbudded calves exhibit measurable increases in anxiety and pessimistic cognitive bias, indicating enduring alterations in affective state akin to those associated with chronic stress in humans. Violence in industrial dairy production is neither exceptional nor accidental; it is embedded in everyday techniques of control that systematically suppress animals' capacities for agency and self-development.

Manuka, a New Zealand-owned dairy company operating in Chile, illustrates this logic at its extremes. In 2014, Chilean authorities launched an investigation into the company – then the largest dairy producer in Latin America – following allegations that approximately 6,000 calves unusable in milk production had been inhumanely killed, with workers reportedly forced to slaughter days-old animals by bludgeoning or air injection and dispose of their bodies in mass graves (Gulliver, 2014). A Chilean politician formally asked the Ministry of Labour to investigate not only the killings but the psychological damage suffered by workers compelled to carry them out. The justification was starkly economic: calves that could not be converted into milk or beef were classified as losses, and eliminating them was deemed rational from the standpoint of profit maximisation. The case reveals the convergence of violence and exploitation at the core of industrial animal agriculture – and demonstrates that the injustice does not hinge on exceptional cruelty. Even where such practices are legally prohibited, the underlying structures of the industry – forced insemination, early calf separation, intensified production cycles, premature slaughter – remain lawful and routine. Law does not merely fail to prevent oppression; it actively legitimises it.

These industries also exemplify cultural imperialism in Young's sense. The consumption of cow's milk – a practice driven by cultural and economic forces rather than biological necessity – has been recast as a universal nutritional norm. Yet the nutritional case is more contested than its cultural dominance suggests. While cow's milk provides nutrients and growth-promoting factors, all are obtainable from other sources, as the historically low dairy intake of many traditional societies demonstrates (Willett & Ludwig, 2020). The primary institutional justification for high dairy consumption – fracture prevention – is far from settled: several large prospective studies find no evidence that higher milk or calcium intake from dairy sources reduces fracture risk in adults (Feskanich et al., 1997; Hegsted, 2001; Willett & Ludwig, 2020), even as other studies report modest protective effects, particularly in certain populations. The nutritional necessity of cow's milk is therefore not an established biological fact but a culturally and institutionally reinforced claim – one whose scientific foundations remain disputed, and whose dominance owes as much to the interests of the dairy industry as to the evidence. This dominant view actively forecloses ethical challenge by naturalising the transformation of animals into instruments of production and rendering their own standpoint invisible. Cultural imperialism here functions not as a separate face of oppression but as the ideological condition that makes the others sustainable: exploitation, marginalisation, powerlessness, and violence are reproduced in part because the dominant culture has made them appear unremarkable.

What emerges from this analysis is not a set of parallel injustices unfolding independently but a single, deeply entangled system of oppression. Migrants are channelled into slaughterhouses and dairy farms precisely because their social and political vulnerabilities render them exploitable; animals are confined within them because their bodies can be transformed into commodities. The same capitalist logic that produces migrant vulnerability at the border organises the systematic exploitation of animal life at the interior. Understanding these industries as spheres of interspecies oppression means recognising that the management of human mobility, the exploitation of migrant labour, and the commodification of animal life are not merely connected by analogy but organised by the same evaluative framework – one that equates value with productivity and treats disposability as the acceptable price of efficiency. To contest one without contesting the other is to leave the underlying structure intact.

6 Conclusion

This chapter has argued that migration is not a neutral or merely descriptive phenomenon, but a locus of oppression in which both human and nonhuman lives are systematically devalued, controlled, and destroyed. By introducing the category of oppressed migrants, I have distinguished forms of mobility compelled by structural injustice from those undertaken under conditions of relative choice. Within this ethical framework, migration appears not as an accident or misfortune, but as the predictable outcome of political, economic, and environmental systems that render life unliveable and then expose those who flee to renewed and intensified harm.

The analysis has further shown that migration cannot be adequately theorised in anthropocentric terms. Border regimes, labour markets, and state policies that oppress human migrants simultaneously entangle and harm nonhuman animals. Fences and walls fracture ecosystems and endanger species; migrant labour in slaughterhouses and farms sustains industries that commodify animal lives. These sites function as spheres of interspecies oppression, in which humans and animals are rendered disposable under the same evaluative regime of productivity and profit maximisation.

Migration also emerges as a multistage process of oppression: beginning in the structural violence that compels departure, unfolding through perilous journeys, intensifying at borders, and persisting in long-term marginalisation within host societies. These stages are analytically distinct but structurally unified. The same mechanisms that disable human development – legal precarity, exposure to harm, and economic coercion – also erode animal autonomy and survival across the landscapes through which migration unfolds. Tracing these entanglements reveals a broader pattern: injustice toward humans and injustice toward animals are mutually reinforcing expressions of the same governing structures.

The normative implications are therefore twofold. First, migration must be understood through the lens of oppression rather than legality or logistics alone. Second, any adequate ethical response to migration must take seriously the interdependence of human and animal fates. Frameworks that treat animal harm as incidental or that normalise disposability across species are normatively insufficient. A just account of migration requires an interspecies account of justice.

This conclusion opens directly onto the next stage of the book's argument. Migration and global extractivism are not only empirically connected but structurally co-produced: the same expansion of industrial commodity chains that generates displacement also destroys the ecosystems and communities – human and nonhuman – that stand in its path. Chapter 5 examines this logic through the lens of global extractivism, tracing how industries such as palm oil production simultaneously displace Indigenous human communities and destroy the habitats and lives of the animals whose world is consumed by the same global commodity chains.

Notes

1 For a contemporary illustration of stages two and three vulnerability, consider current enforcement practices of US Immigration and Customs Enforcement (ICE), including prolonged detention, family separation, restricted access to legal counsel, and exposure to physical and psychological harm. These practices have been widely documented by human rights organisations and courts, and exemplify how border governance systematically produces vulnerability even after migrants have reached the territory of a receiving state.

2 A classic literary illustration of this dynamic is provided by John Steinbeck's *The Grapes of Wrath*, which traces the migration of Dust Bowl families whose movement initially appears economically motivated but is later revealed as exposure to systematic exploitation, coercion, and dispossession within agricultural labour regimes. The novel is useful as a cultural articulation of how migration can shift from an apparent choice to a structurally imposed vulnerability.

3 The construction of Israel's separation wall cannot be analysed in isolation from the broader legal and political context of the occupied Palestinian territory. In its 2004 Advisory Opinion on the Legal Consequences of the Construction of a Wall in the Occupied Palestinian Territory, the International Court of Justice concluded that the wall, insofar as it is built within occupied territory, violates international humanitarian law and international human rights law, including obligations under the Fourth Geneva Convention (International Court of Justice, 2004). Following the attacks carried out by Hamas on 7 October 2023, which resulted in the killing of Israeli civilians and the taking of hostages, Israel launched large-scale military operations in the Gaza Strip. These operations have caused extensive civilian casualties, mass displacement, and the widespread destruction of civilian infrastructure, as documented by United Nations agencies and humanitarian organisations. In this context, the United Nations Independent International Commission of Inquiry on the Occupied Palestinian Territory, including East Jerusalem, and Israel has stated that Israel is responsible for the commission of genocide in Gaza. This broader context underscores how border and separation infrastructures operate within systems of oppression that profoundly affect both human and nonhuman life.

4 For a remarkable first-person account of the psychological and physical toll of a life spent in slaughterhouse labour, see Ponthus, 2019. (English translation exists: *On the*

Line. Notes from a Factory (translated by Stephanie Smee, published by Apollo, 2021.))

5 Although animals in industrial agriculture may exhibit forms of resistance, the institutional conditions under which they are confined deny them any meaningful capacity for refusal (Hribal, 2010). See also Meijer, 2019.

References

Adcock, F. et al. (2015, August). The economic impacts of immigrant labor on U.S. dairy farms. *National Milk Producers Federation*. www.nmpf.org/wp-content/uploads/immigration-survey-090915.pdf.

Aldridge, R. et al. (2018). Global patterns of mortality in international migrants: A systematic review and meta-analysis. *The Lancet*, 392(10164), 2553–2566. doi:10.1016/s0140-6736(18)32781–32788.

Bar-On, Y. M. et al. (2018). The biomass distribution on Earth. *Proceedings of the National Academy of Sciences*, 115(25), 6506–6511. doi:10.1073/pnas.1711842115.

BBC News. (2019, 19 December). Confessions of a slaughterhouse worker. www.bbc.com/news/stories-50986683.

Birke, P., & Bluhm, F. (2020). Migrant labour and workers' struggles: The German meatpacking industry as contested terrain. *Global Labour Journal*, 11(1). doi:10.15173/glj.v11i1.3875.

Bragg, B., & Hyndman, J. (2024). "There is no safe place in this plant": Refugee workers in Canadian meatpacking and the limits of permanent legal status. *Refuge: Canada's Journal on Refugees*. doi:10.25071/1920-7336.41131.

De Angelo, C. et al. (2011). Differential impact of landscape transformation on pumas and jaguars in the Upper Paraná Atlantic Forest. *Diversity and Distributions*, 17(3), 422–436. doi:10.1111/j.1472-4642.2011.00746.x.

EFFAT/European Federation of Food, Agriculture and Tourism Trade Unions. (2020). *COVID-19 outbreaks in slaughterhouses and meat processing plants*. https://effat.org/wp-content/uploads/2020/06/EFFAT-Report-Covid-19-outbreaks-in-slaughterhouses-and-meat-packing-plants-State-of-affairs-and-proposals-for-policy-action-at-EU-level-30.06.2020.pdf.

Feskanich, D. et al. (1997). Milk, dietary calcium, and bone fractures in women: A 12-year prospective study. *American Journal of Public Health*, 87(6), 992–997. doi:10.2105/AJPH.87.6.992.

Fremstad, S. et al. (2020, April 29). Meatpacking workers are a diverse group who need better protections. *Center for Economic and Policy Research*. https://cepr.net/publications/meatpacking-workers-are-a-diverse-group-who-need-better-protections/.

Frydenlund, S., & Dunn, E. (2022). Refugees and racial capitalism: Meatpacking and the primitive accumulation of labor. *Political Geography*, 93. doi:10.1016/j.polgeo.2021.102575.

Gottlieb, N. et al. (2025). Immigrant workers in the meat industry during COVID-19: Comparing governmental protection in Germany, the Netherlands, and the USA. *Globalization and Health*, 21. doi:10.1186/s12992-025-01104-9.

Gottlieb, N., & Holmes, S. M. (2022). Labor migration, the food supply chain and the COVID-19 syndemic: Germany, Netherlands, and the USA [Workshop abstract]. *European Journal of Public Health*, 32(Suppl. 3), ckac129. 404. doi:10.1093/eurpub/ckac129.404.

Government of India. (2021). Border fencing. *Press Information Bureau, Ministry of Home Affairs*. www.pib.gov.in/Pressreleaseshare.aspx?PRID=1741906&reg=3&lang=2.

Greenwald, N. et al. (2017). A wall in the wild: The disastrous impacts of Trump's border wall on wildlife. *Center for Biological Diversity*. www.biologicaldiversity.org/programs/international/borderlands_and_boundary_waters/pdfs/A_Wall_in_the_Wild.pdf.

Gulliver, A. (2014, 14 March). Manuka dairy firm under investigation in Chile. *Stuff*. www.stuff.co.nz/business/farming/dairy/9634584/Manuka-dairy-firm-under-investigation-in-Chile.

Harrity, E. et al. (2024). USA-Mexico border wall impedes wildlife movement. *Frontiers in Ecology and Evolution*, 12. doi:10.3389/fevo.2024.1487911.

Hegsted, D. M. (2001). Fractures, calcium, and the modern diet. *American Journal of Clinical Nutrition*, 74(5), 571–573. doi:10.1093/ajcn/74.5.571.

Hribal, J. (2010). *Fear of the animal planet: The hidden history of animal resistance*. CounterPunch/AK Press.

International Court of Justice. (2004). Legal consequences of the construction of a wall in the occupied Palestinian territory: Advisory opinion. www.icj-cij.org/case/131.

IUCN. (2020). *The IUCN red list of threatened species. International Union for Conservation of Nature*. www.iucnredlist.org.

Kerwin, D., & Martínez, D. (2024). Forced migration, deterrence, and solutions to the non-natural disaster of migrant deaths along the US-Mexico border and beyond. *Journal on Migration and Human Security*, 12(2), 127–159. doi:10.1177/23315024241277532.

Lever, J., & Milbourne, P. (2016). The structural invisibility of outsiders: The role of migrant labour in the meat-processing industry. *Sociology*, 51(2), 306–322. doi:10.1177/0038038515616354.

Linnell, J. D. C. et al. (2016). Border security fencing and wildlife: The end of the transboundary paradigm in Eurasia? *PLOS Biology*, 14(6), e1002483. doi:10.1371/journal.pbio.1002483.

Luxury Pet Source. (n.d.). Pet assurance. https://luxurypetsource.com/pet-assurance/.

McCain, E. B., & Childs, J. L. (2008). Evidence of resident jaguars in the southwestern United States and the implications for conservation. *Journal of Mammalogy*, 89(1), 1–10. doi:10.1644/07-MAMM-A-014.1.

Meijer, E. (2019). *When animals speak: Toward an interspecies democracy*. NYU Press.

Miyake, K. K. (2023). Carceral jaguar geographies along the US/México border and the case for border abolition. *Local Environment*, 28(10), 1295–1310. doi:10.1080/13549839.2023.2206641.

Neave, H. W. et al. (2013). Pain and pessimism: Dairy calves exhibit negative judgement bias following hot-iron disbudding. *PLOS ONE*, 8(12), e80556. doi:10.1371/journal.pone.0080556.

Pachirat, T. (2011). *Every twelve seconds: Industrialized slaughter and the politics of sight*. Yale University Press.

Passarelli-Araújo, H. (2025). Deaths across borders: Trends and challenges in estimating mortality among undocumented migrants on Latin American routes. *Journal of Epidemiology and Population Health*, 73(4), 203125. doi:10.1016/j.jeph.2025.203125.

Peters, R. L. et al. (2018). Nature divided, scientists united: US–Mexico border wall threatens biodiversity and binational conservation. *BioScience*, 68(10), 740–743. doi:10.1093/biosci/biy063.

Ponthus, J. (2019). *À la ligne: Feuillets d'usine*. Éditions de la Table ronde.

Poole, D. et al. (2020). Spatial–temporal trends in forced migrant mortality, 2014–2018. *BMJ Global Health*, 5(7). doi:10.1136/bmjgh-2020-002885.

Reed-Sandoval, A. (2019). The injustice of the 'migrant journey' to the United States. *Critical Review of International Social and Political Philosophy*, 22(6), 747–769. doi:10.1080/13698230.2018.1426817.

Roques, S. et al. (2016). Effects of habitat deterioration on the population genetics and conservation of the jaguar. *Conservation Genetics*, 17, 125–139. doi:10.1007/s10592-015-0766-5.

Shaffer, L. J. et al. (2019). Human-elephant conflict: A review of current management strategies and future directions. *Frontiers in Ecology and Evolution*, 6. doi:10.3389/fevo.2018.00235.

Sironi, A. et al. (Eds.). (2019). *Glossary on migration* (3rd ed.). International Organization for Migration.

Titley, M. A. et al. (2021). Global inequities and political borders threaten conservation of transboundary species. *Proceedings of the National Academy of Sciences*, 118(7), e2011204118. doi:10.1073/pnas.2011204118.

Willett, W., & Ludwig, D. S. (2020). Milk and health. *New England Journal of Medicine*, 382(7), 644–654. doi:10.1056/NEJMra1903547.

Yom-Tov, Y. et al. (2021). The plight of the endangered mountain gazelle Gazella gazella. *Oryx*, 55(5), 771–778. doi:10.1017/S003060531900108X.

Young, I. M. (1990). *Justice and the politics of difference*. Princeton University Press.

5

INTERSPECIES GLOBAL INJUSTICE AND STRUCTURAL RESPONSIBILITY

1 Introduction

Oppression rarely respects boundaries. Those who benefit from injustice are often geographically, socially, and institutionally distant from those who bear its costs. Consumers in one part of the world may unknowingly sustain labour exploitation, ecological devastation, and animal suffering in another. Distance, however, does not dissolve responsibility. On the contrary, it reveals that oppression is organised through transnational structures that bind together disparate lives and landscapes.

In the global sphere, harms cross borders through institutional architectures that fragment agency and obscure accountability. Trade regimes, supply chains, investment agreements, and financial infrastructures coordinate extraction across continents while dispersing responsibility among corporations, subcontractors, regulatory authorities, and consumer markets. Jurisdictional boundaries shield actors from liability even as economic integration binds their actions together. In this setting, injustice seldom appears as direct subordination by identifiable perpetrators; it is embedded in routine economic practices that normalise dispossession and destruction as conditions of prosperity.

Dominant paradigms of global justice are ill-equipped to grasp this structural character. Statist approaches ground principles of justice within bounded political communities, underestimating the coercive force of transnational economic arrangements that shape life chances beyond any single state. Cosmopolitan theories extend distributive principles across borders but typically frame injustice as maldistribution – focusing on allocation rather than on the institutional organisation of production and extraction that generates vulnerability in the first place. What recedes from view in both is a prior question:

DOI: 10.4324/9781003434856-6

not who has how much, but how global structures determine whose lives are secured and whose are rendered disposable.

This chapter reconstructs global justice as a matter of structural responsibility. Drawing on Iris Marion Young's social connection model (Young, 2011), it argues that political responsibility is grounded in participation in the institutional processes that reproduce the background conditions sustaining structural harm across borders. It also advances a second, less familiar intervention: extending this structural account across species lines. Global capitalism does not merely generate inequality among humans; it produces interspecies hierarchies embedded in transnational regimes of production. Animal suffering, habitat destruction, and the commodification of nonhuman life are not peripheral to global capitalism's operation – they are constitutive of how global wealth is made. Interspecies injustice is not an accidental by-product of otherwise neutral systems; it is woven into their normal functioning.

The Malaysian palm oil industry illustrates how these dynamics converge in practice. Palm oil production links multinational corporations, consumer markets, migrant workers, Indigenous communities, and nonhuman animals within a single transnational chain of value extraction, making visible the global entanglement of extractive economies, labour regimes, and ecological devastation. From the forced sedentarisation of the Batek to the systematic displacement and killing of Bornean elephants, a common logic persists: the concentration of privilege through the systematic externalisation of vulnerability.

2 Global Structural Injustice and Responsibility

Debates about global justice are often structured around a familiar divide between cosmopolitan and statist positions. Cosmopolitans argue that because our actions predictably affect people beyond our borders, principles of justice cannot be confined to territorial membership (Beitz, 1979; Pogge, 2002). Moral obligations extend to all persons in virtue of their equal moral status. Peter Singer's influential argument from preventable suffering contends that distance is morally irrelevant: if we can prevent serious harm at modest cost, we are required to do so (Singer, 1972). Simon Caney and other cosmopolitan theorists similarly maintain that global institutions and economic interdependence generate duties of distributive justice that apply universally (Caney, 2005). In this view, global inequality is unjust because it reflects an unfair distribution of resources, opportunities, or welfare among persons who are moral equals.

Statists, by contrast, maintain that principles of justice arise within bounded political communities (Nagel, 2005; Miller, 2007). Because justice regulates coercive institutions, and because such institutions are typically organised at the level of the state, demanding obligations of justice apply primarily among co-citizens. Duties to distant others may exist, but they are typically framed as humanitarian assistance rather than as strict requirements of justice. The state remains the primary site within which egalitarian principles properly operate.

Both positions illuminate important aspects of global moral life. Cosmopolitanism rightly insists that borders do not erase moral concern. Statism underscores the role of shared political institutions in generating special obligations. Yet both approaches risk misidentifying the structure of global injustice. Cosmopolitan theories often frame injustice primarily in distributive terms: who has how much, and how goods should be allocated. Statist theories focus on the locus of coercive authority and the boundaries of political membership. What tends to recede from view in both is the institutional organisation of production, extraction, and exchange that systematically generates vulnerability across borders.

The central question is not only whether goods are distributed fairly, nor merely where coercive authority resides, but how global economic and political structures shape life chances in the first place. Transnational supply chains, trade regimes, financial institutions, and development policies structure relations of dependence and hierarchy that cannot be reduced either to maldistribution or domestic coercion. Injustice emerges from the patterned operation of these processes, not simply from the failure to transfer sufficient resources.

Iris Marion Young rejected the opposition between cosmopolitan universalism and statist restriction. She denied that obligations of justice are grounded solely in abstract shared humanity, but equally rejected the idea that justice stops at the border. Instead, she developed a structural account of global justice centred on the existence of transnational institutions and social processes that link individuals across boundaries. In her view, social relations are not restricted to nation-state borders: our actions are conditioned by and contribute to institutions that affect distant others, and their actions contribute to the function of institutions that affect us (Young, 2007, p. 163).

From this perspective, injustice is not primarily the result of discrete wrongdoing by identifiable perpetrators, but of structural processes to which many contribute and from which many benefit. Young's social connection model highlights how ordinary actions – shopping for clothes, investing money, consuming food – are embedded in institutional arrangements that produce foreseeable harms for distant others. As opposed to legal or moral responsibility, political responsibility cannot be reduced to guilt or blame. It must instead be understood, Young argues, as a forward-looking obligation to participate in the transformation of unjust structures.

Young's account is significant because it reveals how global institutions bind distant lives into durable webs of hierarchy and privilege. Those positioned advantageously within global commodity chains benefit from cheap goods, secure employment, and economic growth, while the social and ecological costs of these advantages are displaced onto structurally disadvantaged populations. The diffusion of agency across networks of production enables beneficiaries to deflect accountability by pointing to proximate agents of harm – factory managers, local officials – while disregarding the wider institutional systems that sustain exploitation.

The limits of a purely distributive approach to global justice become clearer in this light. Cosmopolitan theories rightly reject the moral arbitrariness of borders and insist on the equal moral status of all persons. Yet they frame the central injustice as maldistribution and propose redistribution as the remedy. What this risks obscuring is a prior question: how are the goods to be redistributed produced, and through what social relations? Distribution presupposes production; before goods can be allocated, they must be generated within institutional arrangements that determine whose labour is used, whose land is rendered available, and who – humans and nonhumans alike – is rendered expendable. In other words, the injustice lies not only in who ends up with less, but in the institutional relations that position some as sources of extractable value for others.

It is possible, at least in principle, to imagine a globally redistributive order that continues to rely on intensive animal agriculture, deforestation, and extractive mining, provided that the proceeds are more evenly shared among humans. Such a world might satisfy certain cosmopolitan egalitarian criteria while reproducing forms of interspecies subordination and ecological destruction that remain structurally embedded. The problem is not that distributive principles are misguided but that they are incomplete – and that incompleteness becomes particularly visible once nonhuman animals are brought into view.

Young's social connection model provides the conceptual resources to precisely articulate this structural dimension of global injustice. Young illustrates these dynamics through the case of sweatshops: labour-intensive production sites, often located in the Global South, that form the terminal nodes of transnational supply chains. As Mirjam Müller observes, sweatshops are marked by "excessively low wages, long hours and overtime, health and safety hazards, and insecure contracts and systematic constraints on collective bargaining and unionization" (Müller, 2024, p. 19). These conditions are not anomalous deviations from otherwise fair markets. They reflect broader global capitalist dynamics of outsourcing and subcontracting that enable corporations to externalise labour costs while maintaining control over production.

Such dynamics are historically sedimented. Drawing on Charles Mills's *The Racial Contract*, M. Müller rightly underscores that "slavery, colonial resource exploitation, the destruction of institutions, and the disruption of communities not only contributed to the economic rise of Europe, but also had a profound impact on the development of colonized countries that still structures the options of workers today" (Müller, 2024, p. 25). Sweatshop labour should therefore not be understood merely as the result of isolated corporate misconduct. It is embedded in a global order shaped by colonial extractivism and racialised hierarchies whose legacies continue to structure labour markets and constrain political alternatives.

Inspired in part by the anti-sweatshop movement, Young sought to explain why consumers in affluent societies often experience themselves as implicated in distant labour exploitation, even when their individual purchases are neither illegal nor directly blameworthy. The intuition animating these movements is that participation in ordinary economic practices can generate a form of responsibility that falls short of guilt or legal liability. Consumers do not intend the harms, nor do they directly impose them, yet their patterned participation helps sustain the institutional processes through which those harms are produced.

For Young, all those connected to structural injustice share political responsibility – distinct from moral blame or legal culpability – to join in collective efforts to transform the unjust structures that bind them together. Political responsibility arises not from having wrongfully caused harm, but from participating in social processes that systematically produce it (Young, 2007, pp. 374, 377; Young, 2011, p. 161). The appeal of the social connection model lies precisely in its capacity to make sense of this intermediate form of responsibility. It explains how individuals may be implicated in injustice without being individually guilty, and why structural harms demand collective political action rather than the identification of isolated perpetrators.[1]

Yet Young leaves under-specified a crucial element of the theory: the nature of connection itself. She does not offer a systematic account of what kinds of relations generate political responsibility, nor how degrees of connection might shape the scope or weight of such responsibility. Is mere participation in global markets sufficient? Does benefit from unjust structures matter more than causal contribution (Young, 2007, pp. 371, 382; Young, 2011, p. 161)? How should we assess responsibility when connection is mediated through complex institutional layers that obscure both agency and impact? These questions become especially pressing in the global context, where supply chains stretch across continents, and responsibility is dispersed across innumerable actors.

Because Young's model assigns political responsibility on the basis of connection, its normative force depends on clarifying what that connection consists in. Without further specification, the theory risks either overextension – implicating everyone equally in all structural injustice – or indeterminacy, offering little guidance about how responsibility should be distributed in practice.

A more precise account is therefore required.

A coherent and consistent way to understand connection to structural injustice is in terms of the reproduction of background conditions. Individuals are connected to structural injustice insofar as their ordinary actions help sustain and reproduce the institutional and material conditions within which they themselves act. Through consumption, investment, labour, political participation, and compliance with prevailing economic norms, agents sustain the very systems that generate patterned vulnerability. Political responsibility arises not because individuals directly cause specific harms, nor merely because they benefit from unjust arrangements, but because their ongoing participation contributes to maintaining the structural environment that renders such harms predictable and systematic.

In this view, connection is neither reducible to causal proximity nor exhausted by distributive benefit. It concerns one's role in sustaining the background conditions of social life. Young gestures toward this idea when she speaks of participation in structural processes, but the notion requires further elaboration. Understanding connection as the reproduction of background conditions provides a principled basis for assigning political responsibility without collapsing it into guilt or blame. Responsibility tracks structural positioning and participation rather than isolated acts of wrongdoing.

Conceived in this way, the social connection model gains normative traction in the global sphere. Political responsibility attaches not because individuals directly intend or cause specific harms, but because their routine participation helps stabilise the institutional environment within which injustice is systematically produced. The significance of sweatshops – Young's own paradigm case of transnational structural injustice – lies not only in their exploitative conditions but in the diffusion of responsibility across transnational networks. Consumers in Europe or North America frequently attribute blame exclusively to local managers or subcontractors, attributions that obscure the wider arrangements: corporate law, trade agreements, and financial flows that connect purchasing power in affluent markets to labour vulnerability elsewhere. In this way, those who are privileged appear detached from the harms that sustain their advantages, even as their ordinary practices reproduce the background conditions under which those harms persist.

This insight extends beyond labour exploitation. The same global structures that link consumers to distant workers also bind human and nonhuman lives within shared regimes of production. Yet even structural accounts of global injustice have largely remained anthropocentric: animals and ecosystems appear as background conditions, economic resources, or collateral damage rather than as subjects implicated in relations of justice. This framing obscures the extent to which global economic systems systematically organise animal bodies and habitats as sites of extraction. Industrial agriculture, deforestation, wildlife commodification, and resource mining are not peripheral to global capitalism's operation – they are central to it. The palm oil industry in Malaysia makes this visible with particular force.

3 From Soy to Palm Oil: Colonial Continuities

Eva Meijer (2023) develops a multispecies extension of Young's social connection model that is particularly instructive for the argument that follows. Applying the framework to interspecies injustice, she examines the case of Dutch animal farmers whose demand for cheap soy links them directly to deforestation in the Amazon and the dispossession of Indigenous communities.

The Netherlands, the second-largest importer of Brazilian soy, relies heavily on this crop to feed animals in industrial agriculture – more than 95 per cent of Brazil's soy production is used for animal feed (Alexandratos & Bruinsma, 2012). This transnational chain of production, driven by European consumption and structured by global trade regimes, has devastating consequences for biodiversity, Indigenous livelihoods, and nonhuman animals in the Amazon.

What Meijer's example makes visible is a structural logic continuous with the dynamics identified in sweatshops: cost minimisation, externalisation of risk, and the displacement of vulnerability operate across species lines. The networks Young describes do not bind together human agents alone. The same supply chains that connect consumers to distant workers also connect them to confined animals, cleared forests, polluted rivers, and displaced wildlife. Political responsibility, grounded in participation in structures that systematically generate harm, cannot be confined to human victims – and global institutions do not merely distribute goods among humans; they organise multispecies life through regimes of extraction, commodification, and disposability.

Palm oil production in Southeast Asia deepens this picture. Now the most widely used vegetable oil in the world – found in processed foods, cosmetics, and biofuels alike (FAOSTAT, 2023; Corley & Tinker, 2016) – palm oil's global dominance is inseparable from its colonial origins. Malaysia, the second-largest global producer, developed its industry within a colonial political economy. Oil palms, not native to the region, were introduced in the nineteenth century by British and Dutch colonial actors. As Jonathan Robins shows, plantation agriculture in the region was embedded in systems of land appropriation, racialised labour control, and what colonial administrators described as rational land use (Robins, 2020; 2021). Like rubber, extracted from the Amazon and transplanted to Southeast Asia, oil palms became instruments of large-scale extraction oriented toward external markets.

This history is not merely background. It illuminates the structural continuity between colonial extraction and contemporary global production. The logics of land alienation, labour discipline, and export-oriented monoculture remain inscribed in the present organisation of the palm oil industry. The environmental consequences are profound: large-scale deforestation, biodiversity loss, soil erosion, and substantial greenhouse gas emissions (Ayompe et al., 2021; Szulczyk & Khan, 2018). Entire ecosystems are cleared and reorganised into monocultural landscapes designed for export efficiency rather than ecological resilience. At the same time, the industry is associated with persistent social injustices, including land dispossession, corruption, unsafe labour conditions, intimidation of activists, and gender-based violence (Mei et al., 2022).

Child labour remains deeply entrenched within this system. Stateless, migrant, and Indigenous children are compelled to perform hazardous plantation work under exploitative conditions – exposed to agrochemicals, heavy physical labour, and chronic exhaustion – while excluded from formal education. A 2018 Malaysian government survey estimated that approximately 33,600 children aged five to seventeen were working in the oil palm industry, with nearly 60 per cent of them located in Sabah alone (ILO, 2020). In that region, families often avoid sending children to alternative schools for fear of deportation, trapping children within cycles of precarity and invisibility that are continuous with the dispossession visited upon Indigenous communities whose territories the industry has consumed (Wahab & Dollah, 2023).

Taken together, these features reveal how contemporary palm oil production reproduces hierarchies first consolidated under colonial rule. Vast monocultures, precarious labour arrangements, ecological degradation, and the displacement of Indigenous communities are not accidental by-products but structural components of a transnational system oriented toward cost minimisation and export growth. The benefits of this system – cheap consumer goods, biofuel inputs, and corporate profits – accrue primarily to actors positioned advantageously within global markets, while the social and ecological costs are borne by migrant workers, Indigenous populations, and nonhuman animals whose habitats and lives are organised as inputs to the same system. The diffusion of responsibility across transnational networks enables beneficiaries to remain insulated from these consequences, even as their consumption patterns help reproduce the background conditions under which such harms persist.

Defenders of the industry frequently emphasise palm oil's contribution to GDP growth and rural development as offsetting justifications for its social and ecological costs. Yet aggregate economic gains do not neutralise structural harms. Framing the issue in terms of net growth invites precisely the kind of moral abstraction that obscures distributional realities. GDP-based metrics capture increases in national output while remaining largely indifferent to who benefits, who bears the costs, and whether fundamental capabilities are secured (Nussbaum, 2006). To weigh economic expansion against the exploitation of children or the dispossession of Indigenous communities is to treat those affected as instruments of collective prosperity rather than as subjects of justice.

Understood through the lens of structural injustice, palm oil plantations are not merely sites of economic activity but institutional spaces in which multiple forms of oppression converge. Exploitation, marginalisation, powerlessness, cultural imperialism, and violence – Young's five faces of oppression – interact within a system oriented toward profitability and extractive efficiency. These dynamics do not operate episodically or independently; they are mutually reinforcing components of a structural order in which privilege is stabilised by displacing vulnerability onto those positioned as expendable.

4 Cultural Imperialism Across Species

Among Young's five faces of oppression, cultural imperialism warrants particular attention in the context of palm oil production. Oppression here operates not solely through labour exploitation or ecological destruction but through the imposition of external norms that reorganise both symbolic and material life. The expansion of palm oil plantations does not merely clear forests; it reconstitutes landscapes according to global economic imperatives, displacing alternative understandings of land, value, and belonging.

4.1 The Batek

The experience of the Batek, a hunter-gatherer community of approximately 1,800 individuals in peninsular Malaysia, illustrates this dynamic with particular force. For the Batek, the forest is not a commodity or natural resource but a living world that sustains subsistence, social relations, and cosmology (Lye, 2002, 2004). As one Batek elder expressed it, as recorded by Tacey (2013, p. 251), the earth is a body whose skin – the forest – has been stripped away, leaving it sick. This image captures not only environmental degradation but a rupture in the ontological framework through which the Batek understand themselves and their place in the world.

Colonial doctrines of rational land use legitimised the dispossession and sedentarisation of mobile peoples such as the Batek (Doolittle, 2005). Contemporary palm oil expansion inherits and rearticulates these logics. Forest territories are cleared, customary land claims marginalised, and communities progressively incorporated into state-directed settlement schemes and formal education systems (Doolittle, 2005; Lye, 2004). The values embedded in global markets – productivity, efficiency, export growth – are naturalised as universal standards against which Indigenous practices are judged deficient or backward. In this process, the Batek's spiritual and cosmological relationships to the forest are not merely ignored but subordinated to monocultural economic rationalities aligned with state and corporate interests.

This is cultural imperialism in Young's sense: the universalisation of a dominant group's experience and perspective, coupled with the marginalisation or distortion of others. What is at stake is not only territorial displacement but a deeper form of dispossession – the erosion of meaning, belonging, and forms of life that do not conform to prevailing extractive paradigms. Crucially, this transformation is not driven by local prejudice alone. It is embedded within the transnational demand for palm oil and the institutional arrangements that render monoculture plantations economically rational. Cultural imperialism thus emerges as a structural effect of global production, linking distant consumers to the reorganisation of lifeworlds across the territories they touch. For mobile peoples, this reorganisation takes a specific spatial form: forced sedentarisation, the fixing-in-place of communities whose modes of being depend on movement. To settle the Batek is not merely to relocate them but to dismantle the form of life through which they exist as Batek.

Nonhuman animals inhabit these forests too, and their worlds are dismantled by the same processes. The conversion of living territories into monocultural surfaces of extraction – what dismantles the Batek's form of life – also reorders the lives of Bornean elephants, and the countless other beings whose habitats are consumed by the same logic. The next section turns to the Bornean elephant, whose displacement makes visible not only the extension of cultural imperialism to nonhuman beings but also the structural violence and distributive limits that define the global palm oil economy.

4.2 Oppressing Elephants

Elephants in Malaysia – alongside tigers, macaques, and numerous other species – are displaced by the expansion of palm oil monocultures. As forest corridors are fragmented and converted into plantation landscapes, habitat ranges shrink and populations decline. The Bornean elephant, a subspecies of the Asian elephant, is now classified as endangered, with approximately 1,000 in the wild, the majority in Sabah (Williams et al., 2020; Saaban et al., 2011; IUCN, 2024). Habitat conversion, rather than isolated poaching, is the primary driver of this vulnerability.

Displacement is not only ecological but institutional. Wildlife governance frameworks routinely categorise elephants as instances of human–wildlife conflict, thereby positioning them as management problems to be addressed through technical interventions such as fencing, electrification, culling, or translocation (Fernando et al., 2012; Ponnusamy et al., 2016). Within such frameworks, elephants are treated as obstacles to development rather than as inhabitants of territories transformed by plantation expansion. Conflict is framed as a behavioural problem requiring technical management rather than as the predictable outcome of land-use regimes structured around export monoculture (Massei et al., 2010). The marginalisation of elephants thus mirrors the marginalisation of Indigenous communities: both are repositioned as impediments to economic rationality.

This dynamic exemplifies a broader form of interspecies cultural imperialism – extending Young's concept beyond its original human-centred formulation. Cultural imperialism, understood as the erasure or subordination of a form of life rather than merely the distortion of a group's conscious self-understanding, applies with force to elephants: beings with complex migratory routes, intergenerational social bonds, spatial memory, and ecological relationships that constitute a distinctive way of inhabiting the world – one that the developmentalist logic of global palm oil demand renders invisible, irrelevant, and expendable.

Elephants endure not only displacement but also direct and systemic forms of harm that reflect the routine functioning of the palm oil economy. Classified as threats to production, they are electrocuted, shot, poisoned, or killed using heavy machinery and firecrackers (Osborn & Parker, 2003; Torre et al., 2021). These practices are not episodic deviations from an otherwise neutral system. They are institutionalised strategies of control embedded in agro-industrial landscapes structured around monocultural efficiency. Their purpose is straightforward: to eliminate whatever disrupts productivity.

In this respect, elephants are incorporated into the same managerial logic that governs other elements of plantation ecology. Like so-called pests or competing plant species, they are treated as obstacles to be removed to optimise yields. The elimination of elephants is framed as a technical matter of agricultural management rather than as violence – a reframing that transforms lethal interventions into routine practices of conflict mitigation, normalising and obscuring the injustice involved.

Here the concept of structural violence becomes indispensable. Structural violence does not refer only to overt acts of physical harm; it denotes the organisation of economic and social systems in ways that systematically disable, displace, and destroy while presenting these outcomes as necessary or inevitable. For elephants, structural violence operates through electric fencing that restricts migratory routes, the translocation of herds to increasingly marginal territories, and the progressive fragmentation of habitats into disconnected corridors. These measures make clear that elephants have no legitimate place within landscapes designed for export-oriented production. Their displacement and death are not accidents but predictable consequences of a global system that organises land primarily around profitability rather than interspecies coexistence.

Young's framework clarifies the depth of this injustice. The face of violence is starkly visible: elephants are routinely subjected to physical harm and death through practices that are not merely tolerated but institutionalised as legitimate instruments of agricultural management. At the same time, they experience powerlessness in a literal sense – the systematic denial of their agency, autonomy, and capacity to inhabit and shape their environments. Within the palm oil economy, elephants are not recognised as beings with social complexity, ecological embeddedness, and a capacity for grief and mourning; instead, they are rendered institutionally invisible – reduced to obstacles whose continued existence is contingent on compatibility with industrial profitability.

The profits secured through this system do not accrue primarily to those who bear the ecological and social burdens of coexistence with elephants. They are concentrated among corporations, investors, and consumers situated advantageously within global markets, who benefit from inexpensive palm oil in processed foods, cosmetics, and biofuels. The destruction of elephant habitats and the violence inflicted upon them thus function as externalised costs, disproportionately borne by marginalised human communities – such as the Batek – and by elephants themselves. The pattern mirrors a longstanding colonial dynamic: extraction and dispossession occur in the Global South, while the privileges of consumption and profit are enjoyed elsewhere.

Yet the logic of disposability identified here is not confined to the Global South. Industrialised countries sustain internal regimes of commodification through intensive animal agriculture that replicates the same structural features in different settings: extreme confinement, labour precarity, ecological degradation, and the systematic erasure of animal agency. As David Nibert (2013) has argued, such systems constitute a form of domesecration – the transformation of animal life into property through enclosure and control. Whether in Borneo or Iowa, animal lives are reorganised within production regimes that render them invisible and disposable. What differs is geography and scale; what remains constant is the institutional logic through which living beings are converted into units of extractable value. The palm oil case and the factory farm are not exceptions to an otherwise benign global order; they are its normal expression. Recognising this requires a fundamental reorientation of how global justice is theorised, taking seriously the institutional production of disposability rather than merely its distributive consequences.

The elephant case makes this limitation concrete: no redistribution of palm oil profits addresses the institutional logic that renders elephant life expendable in the first place. While indispensable, the distributive paradigm is insufficient for grasping the character of interspecies injustice. Distribution presupposes a fixed set of recognised subjects among whom benefits and burdens are to be apportioned. Yet, as the preceding analysis shows, global injustice also operates by determining who counts as a subject of justice at all.

What is required is not the abandonment of distribution but its reorientation within a relational framework that foregrounds interdependence, embeddedness, and structural entanglement. Understood relationally, global justice demands the transformation of the structures that naturalise harm, not merely the redistribution of its outcomes. It requires reconfiguring the material and symbolic conditions under which some lives are rendered visible and protected while others are confined, erased, or killed. For nonhuman animals, this entails more than prohibiting cruelty. It requires dismantling the institutional infrastructures – supply chains, legal classifications, economic incentives – that sustain their commodification. Global justice, in this view, is inseparable from the transformation of the production systems that tether prosperity to disposability.

5 Conclusion

This chapter has argued that global justice cannot be meaningfully addressed without recognising that structural oppression traverses not only territorial boundaries but also species lines. The pursuit of economic growth and consumer convenience in industrialised societies is inseparable from the production of spaces of dispossession elsewhere – and those spaces are inhabited not only by exploited human workers and displaced Indigenous communities but by nonhuman animals whose habitats and lives are organised as inputs to global commodity chains. These are not aberrations within an otherwise neutral global order. They are structural features of a political economy that binds distant actors into relations of dependency, privilege, and diffuse irresponsibility.

The Malaysian palm oil industry has shown how colonial continuities, cultural imperialism, labour exploitation, and multispecies violence converge within a single transnational system. The forced sedentarisation of the Batek, the exploitation of migrant and child labour, and the systematic displacement and killing of Bornean elephants are not isolated wrongs. They are interconnected expressions of a global structure that organises land, labour, and animal life around profitability and export growth. What appears as development or agricultural management at one site is sustained by patterns of consumption and investment elsewhere – and it is precisely this distance that enables beneficiaries to remain insulated from the harms their participation helps reproduce.

The chapter's central theoretical contribution has been to specify what connection to structural injustice consists in: not proximate causation, not merely benefit, but the reproduction of background conditions – the ongoing participation in institutional processes that sustain the structural environment within which harm is predictably and systematically generated. In this account, responsibility for interspecies oppression is not confined to plantation managers, corporate executives, or state officials. It extends to all those whose ordinary economic practices help stabilise the conditions under which such oppression persists. This is a demanding conclusion, but it follows directly from taking seriously the structural character of global injustice.

Seen alongside the analyses of disability and migration in the preceding chapters, a common logic becomes visible. Hierarchies of worth designate some individuals, communities, and species as productive and valuable while rendering others marginal, dependent, or disposable. These hierarchies are stabilised through institutional arrangements that diffuse responsibility and conceal structural violence. Across labour markets, border regimes, plantations, and factory farms, privilege is secured by externalising vulnerability – and that externalisation operates across species lines with the same systematic force as it operates across class, race, and national boundaries.

* * *

This closes Part I of the book, which has traced how oppression operates across bodies, borders, and species. It has been shown that injustice is embedded in the background conditions of social and economic life, reproduced through ordinary practices that appear routine or inevitable. Critique of these structures is necessary but insufficient. The question that now emerges is institutional and constructive: what would it mean to reorganise political communities around principles of care, interdependence, and multispecies justice rather than hierarchy, extraction, and disposability? Answering this question requires not only a different set of institutions but a different moral orientation – one capable of recognising the vulnerability of human and nonhuman lives alike as a site of political responsibility rather than a problem to be managed. Part II takes up this question. Building on the ethic of care and civic solidarity developed in Chapter 6, it turns from the diagnosis of structural injustice to the design of institutions capable of responding to it (Chapters 7 to 9) – asking how the conditions for self-development might be secured not only for human beings, but across the full range of lives that share our world.

Note

1 For a systematic discussion of this issue, see McKeown 2018.

References

Alexandratos, N., & Bruinsma, J. (2012). World agriculture towards 2030/2050: The 2012 revision. *FAO Agricultural Development Economics Division*. www.fao.org/3/ap106e/ap106e.pdf.

Ayompe, L. M. et al. (2021). Towards sustainable palm oil production: The positive and negative impacts on ecosystem services and human wellbeing. *Journal of Cleaner Production*, 289, 125737. doi:10.1016/j.jclepro.2020.123914.

Beitz, C. (1979). *Political theory and international relations*. Princeton University Press.

Caney, S. (2005). *Justice beyond borders: A global political theory*. Oxford University Press.

Corley, R. H. V., & Tinker, P. B. (2016). *The oil palm* (5th ed.). Wiley-Blackwell.

Doolittle, A. A. (2005). *Property and politics in Sabah, Malaysia: Native struggles over land rights*. University of Washington Press.

FAOSTAT. (2023). Crops and livestock products. *Food and Agriculture Organization of the United Nations*. www.fao.org/faostat/en/#data/QCL.

Fernando, P., et al. (2012). Problem-elephant translocation: Translocating the problem and the elephant? *PLOS ONE*, 7(12), e50917.

International Labour Organization. (ILO) (2020). Global estimates of child labour: Results and trends, 2012–2016. *International Labour Office*. www.ilo.org/global/publications/books/WCMS_575499/lang–en/index.htm.

IUCN. (2024, June 6). Bornean elephant endangered – IUCN Red List. www.iucn.org/press-release/202406/bornean-elephant-endangered-iucn-red-list.

Lye, T.-P. (2002). The significance of forest to the emergence of Batek knowledge in Pahang, Malaysia. *Southeast Asian Studies*, 40(1), 3–38.

Lye, T. P. (2004). *Changing pathways: Forest degradation and the Batek of Pahang, Malaysia*. Lexington Books.

Massei, G. et al. (2010). Can translocations be used to mitigate human–wildlife conflicts? *Wildlife Research*, 37(5), 428–439. doi:10.1071/WR08179.

McKeown, M. (2018). Iris Marion Young's "social connection model" of responsibility: Clarifying the meaning of connection. *Journal of Social Philosophy*, 49, 484–502. doi:10.1111/josp.12253.

Mei, L. et al. (2022). Identifying the human rights impacts of palm oil: Guidance for financial institutions and downstream companies. *Forest Peoples Programme*. https://globalcanopy.org/wp-content/uploads/2022/08/FPP-Palm-Oil-Report-FINAL52.pdf.

Meijer, E. (2023). Global injustice and animals: Towards a multispecies social connection model. *International Relations*, 37(3), 497–513.

Miller, D. (2007). *National responsibility and global justice*. Oxford University Press.

Müller, M. (2024). *Global sweatshops: A feminist theory of exploitation and resistance.* Oxford University Press.

Nagel, T. (2005). The problem of global justice. *Philosophy & Public Affairs*, 33(2), 113–147.

Nibert, D. (2013). *Animal oppression and human violence: Domesecration, capitalism, and global conflict.* Columbia University Press.

Nussbaum, M. C. (2006). *Frontiers of justice: Disability, nationality, species membership.* Harvard University Press.

Osborn, F. V., & Parker, G. E. (2003). Towards an integrated approach for reducing the conflict between elephants and people: A review of current research. *Oryx*, 37(1), 80–84. doi:10.1017/S0030605303000152.

Torre, J. et al. (2021). There will be conflict – agricultural landscapes are prime, rather than marginal, habitats for Asian elephants. *Animal Conservation*, 24. doi:10.1111/acv.12668.

Pogge, T. (2002). *World poverty and human rights.* Polity Press.

Ponnusamy, V., et al. (2016). Farmers' perceptions and attitudes towards government-constructed electric fences in Peninsular Malaysia. *Gajah*, 45, 4–11.

Robins, J. (2020). Shallow roots: The early oil palm industry in Southeast Asia, 1848–1940. *Journal of Southeast Asian Studies*, 51(4), 538–560. doi:10.1017/S0022463420000697.

Robins, J. (2021). *Oil palm: A global history.* University of North Carolina Press.

Saaban, S. et al. (2011). Current status of Asian elephants in Peninsular Malaysia. *Gajah*, 35, 67–75.

Singer, P. (1972). Famine, affluence, and morality. *Philosophy & Public Affairs*, 1(3), 229–243.

Szulczyk, K. R., & Khan, A. R. M. (2018). The potential and environmental ramifications of palm biodiesel: Evidence from Malaysia. *Journal of Cleaner Production*, 203, 260–272. doi:10.1016/j.jclepro.2018.08.258.

Tacey, I. (2013). Tropes of fear: The impact of globalization on Batek religious landscapes. *Religions*, 4(2), 240–266. doi:10.3390/rel4020240.

Wahab, A., & Dollah, R. (2023). Measuring Child Labor in Oil Palm Production in Sabah, Malaysia. *Sage Open*, 13(4).

Williams, C. et al. (2020). *Elephas maximus.* The IUCN Red List of Threatened Species.

Young, I. M. (2007). *Global challenges: War, self-determination and responsibility for justice.* Polity.

Young, I. M. (2011). *Responsibility for justice.* Oxford University Press.

PART II

Toward Interspecies Justice

6

THE FOUNDATIONS OF ANIMAL JUSTICE

Caring Politics and Civic Solidarity

1 Introduction

The first part of the book concluded that animals, like humans, can be subjects of oppression. Drawing on Young's account, I argued that oppression is a structural injustice rooted in group-based hierarchies and characterised by its disabling power – the systematic hindrance of self-development. While the claim that oppression is a specific form of injustice may appear uncontroversial, it carries particular weight in the present context. Not only does it capture the prevailing and most consequential relationship between humans and many other animals in industrialised societies, but taking it seriously also clarifies both the distinctive wrongs animals suffer and the kinds of political responses such injustice demands.

Building on this conclusion, the second part of the book advances the non-speciesist interpretation of Young's account while shifting the focus from diagnosis to prescription. The task is no longer to identify injustice but to ask what values and principles should guide political institutions in a multispecies society structured by relations of power, dependency, and vulnerability. I argue that such institutions must be grounded in principles of care and solidarity, reflecting the ethos of an interspecies politics of care that is attentive not only to need but also to oppression.

To pursue this task, the chapter proceeds in three steps. First, I look at the attitudes that sustain oppressive human–animal relationships and that constitute what I call speciesist privilege. Second, to respond to it, I offer a framework of an interspecies politics of care – one that confronts the inattentiveness embedded in privilege with attention and responsiveness. The framework is developed with particular emphasis on its implications for institutional design. Third, despite its central role in my account of justice, I

DOI: 10.4324/9781003434856-8

contend that care alone is insufficient to dismantle entrenched structures of privilege, and I therefore supplement the notion of caring institutions with the principle of solidarity. Solidarity, as I understand it here, is directed at all those who can be or are being oppressed within a political community, expressing an institutional commitment to challenging the structural conditions that enable oppression. The combination of care and solidarity, I argue, is necessary to underpin just political institutions in a multispecies society.

2 Privilege and Oppression

As the preceding analysis has shown, oppression does not merely involve the infliction of harm but the organisation of the social world through hierarchical distinctions that rank beings and normalise asymmetries of power. Those positioned higher in such hierarchies enjoy systematic advantages that accrue not merely alongside, but at the expense of, those positioned below.

This hierarchical organisation of society directs our attention to the concept of privilege: the benefits that dominant groups, and the individuals who constitute them, enjoy because social structures are arranged in their favour. Privilege is not reducible to a single dimension. It encompasses material advantages – access to goods, services, and opportunities produced through the exploitation of subordinated groups; epistemic advantages – the socially supported capacity to remain ignorant of the costs and harms on which those benefits depend; and political advantages – the institutional protection of those benefits through law, policy, and cultural norms. It is the intersection of these dimensions that gives privilege its particular resilience, a resilience sustained in part by the denial of those who benefit from it – by their disposition to obscure or downplay the very structures that sustain their advantage. This denial is one of the mechanisms through which oppression reproduces itself.

2.1 Privileged Irresponsibility

Political theorist Joan Tronto introduced the concept of privileged irresponsibility in the early 1990s to capture a central normative mechanism through which structural injustice is sustained (Tronto, 1990; 1993). The term refers to how those occupying advantaged social positions fail to recognise their dependence on the care, labour, and vulnerability of others, treating these supports instead as natural entitlements flowing from their social status. Caregivers are thus instrumentalised to sustain privilege, while their contributions remain socially and politically invisible. This normalisation of advantage renders the privileged either unable or unwilling to see that their position within the political community is sustained by relations of power rather than by merit or inherent worth. The result is a cultivated inattentiveness to the needs and claims of the unprivileged – an inattentiveness that not only perpetuates structural injustice but helps to conceal it (Tronto, 1993; Bozalek, 2014).

Although Tronto develops the notion of privileged irresponsibility partly through an analysis of racial injustice, she explicitly connects this stance to the exercise of power more generally.[1] In the context of racism, Tronto observes that oppressors tend to forget the wrongs they have committed, while those who were oppressed still find it difficult to free themselves – a selective memory that sustains the self-exonerating claim that "we all have the same choices now" (Tronto, 2013, p. 128). Those who benefit from racial hierarchy fail to recognise that their advantages are sustained by social structures that impose constraints and harms on others. Privileged irresponsibility thus names a stance through which oppression is rendered ordinary and morally unremarkable to those who benefit from it.

Tronto shows that comparable dynamics operate across a range of structural injustices, including patriarchy, ableism, and heteronormativity. In each case, social hierarchies are stabilised through the normalisation of advantage and the systematic rendering of dependence invisible. Certain groups are instrumentalised – made to bear the unacknowledged costs of institutional arrangements – in order to secure and reproduce the privileged positions of others. Privileged irresponsibility therefore emerges as a general structural mechanism through which oppression is reproduced while remaining largely unseen by those who benefit from it.

Understood in this way, privileged irresponsibility operates on two levels at once. It is a general structural mechanism through which oppression is reproduced; but it is also a set of individual and collective dispositions through which that mechanism is enacted in everyday life. Oppression persists not only through coercion or exclusion but through privileged dispositions that normalise unjust relations, rendering them acceptable, justified, or even invisible. Confronting privileged irresponsibility is therefore a requirement of any politics committed to justice, exposing the relations of power it conceals and challenging the legitimacy of the advantages it normalises.

2.2 Privileged Ignorance

While privileged irresponsibility names a normative–structural stance toward others, this stance is sustained and reinforced by a distinct but closely related epistemic dimension: privileged ignorance. That stance – marked by inattentiveness, entitlement, and denial – depends on a socially supported capacity to avoid the knowledge that would make it untenable. This is not a matter of mere forgetfulness or accidental lack of information. When Tronto describes privileged irresponsibility as involving selective memory – where oppressors forget the wrongs they did – she gestures toward an underlying epistemic orientation that systematically shields the privileged from unsettling knowledge.

This epistemic dimension of privilege has been developed most explicitly by José Medina as active ignorance (Medina, 2013; see also Bowman, 2020). This is not a passive absence of knowledge but an epistemic vice: a socially supported disposition that enables the advantaged to avoid the cognitive and moral dissonance that would follow from recognising the injustices underpinning their social position. As Medina puts it, active ignorance is "the privilege of not knowing or of not needing to know" (Medina, 2013, p. 32).

Adopting part of Medina's analysis, I trace privileged ignorance across a wide range of social contexts. In many societies, men have not been expected to understand or participate in domestic and reproductive labour, despite their dependence on it, because patriarchal hierarchies insulated them from the practical and moral demands of such knowledge. Similarly, within authoritarian regimes, those who benefit from dictatorship often enjoy the luxury of ignoring the violence inflicted on fellow citizens. This was evident in Pinochet's Chile, where many supporters of the regime later claimed ignorance of its atrocities, insisting that they would not have supported the dictator had they known.[2]

Such claims exemplify the privilege of not having, or not needing, to know. The issue is not a lack of access to information but the absence of social and moral pressure to attend to it. As Medina puts it, "despite their greater access to the facts of social life, privileged subjects often ignore the most violent and hard-to-swallow aspects of social confrontation. Those with enormous amounts of power can sometimes be quite ignorant of the violence committed to maintain their privilege, and can be genuinely surprised when they encounter facts about the oppression of which they are the main beneficiaries" (Medina, 2013, p. 33).

Oppression, then, is sustained not only through material hierarchies and institutional arrangements but also through epistemic structures that protect the privileged from recognising their own complicity. Privileged ignorance reinforces privileged irresponsibility with a cultivated unawareness, enabling oppression to persist without sustained political challenge.

2.3 Speciesist Privilege

With these two mechanisms in view, we are now in a position to address the question of animals. Since, as argued in Part I, animal oppression is a structural feature of industrialised societies, what kind of privilege is operative in our relationships with other animals?

The answer lies in speciesist privilege, which emerges precisely at the intersection of these normative and epistemic dynamics. Speciesist privilege refers to the systematic tendency of humans to treat the benefits they derive from animals – food, labour, companionship, scientific use, and entertainment – not as outcomes of power-laden social arrangements but as natural entitlements beyond moral scrutiny (see Van den Brandeler, 2024).

A particularly revealing site for observing this intersection is industrial animal agriculture. Here, privileged irresponsibility and privileged ignorance operate together to normalise extreme forms of injustice. Through processes of invisibilisation and cultivated forgetfulness, humans transform domesticated animals into permanent victims of oppression while securing the routine satisfaction of human interests with minimal cognitive or moral dissonance. In this context, privilege does not appear as an exception or moral failure but as an ordinary condition of everyday social and economic life.

This dynamic is vividly illustrated in the meat, dairy, and egg industries, where animals' bodies are instrumentalised, their suffering rendered invisible, and their existence reduced to disposable resources.[3] Timothy Pachirat's ethnography *Every Twelve Seconds* (2011) provides a particularly illuminating account of how this invisibilisation is institutionally produced and maintained.

The slaughterhouse, in Pachirat's analysis, is not merely a workplace, but a system of concealment. Its architecture, its compartmentalisation of labour, and the spatial and psychological distance between killing and consumption function together to obscure violence against animals. These arrangements enable consumers to benefit from the system without confronting its moral costs, while also shielding many workers from full awareness of the collective harm their labour sustains. Ignorance here is not accidental but structurally organised.

What this analysis makes clear is that speciesist privilege is not reducible to individual attitudes or moral blind spots. It is actively reproduced through institutional design. By concealing the violence on which it depends, industrial animal agriculture naturalises the benefits humans derive from animals, allowing these advantages to appear legitimate, inevitable, and morally unproblematic. Privileged irresponsibility and privileged ignorance are thus woven into the fabric of human–animal relations in industrialised societies.

The invisibility that Pachirat describes is not only architectural and psychological, but also legal. Across different jurisdictions, so-called ag-gag laws regulate what can – and crucially, cannot – be made visible about the lives of animals in industrial farming (Ceryes & Heaney, 2019). By criminalising documentation and targeting whistle-blowers, journalists, and activists, these laws do more than protect economic interests: they institutionalise ignorance. Structural violence against animals is shielded from public scrutiny, preserving the appearance of legitimacy in practices that would otherwise provoke moral and political contestation (Wilson, 2014; McCausland et al., 2013; Robbins et al., 2016).

Underlying both mechanisms is a long-standing hierarchical worldview that gives them their cultural durability: the tradition of the *scala naturae*. As Martha Nussbaum has observed, many Euro-American cultures have absorbed an image of nature as a ladder, with humans occupying the highest rung (Nussbaum, 2023, p. 20). This conception functions not merely as a metaphysical doctrine but as an organising framework for social life, shaping food systems, entertainment, environmental governance, education, and political standing.

These hierarchies are reproduced intergenerationally through institutional practices such as schooling. Curricula that present animals alongside minerals as "natural resources," and school food programmes that normalise animal consumption, habituate children into a worldview in which animals appear as consumable objects rather than as beings with interests of their own. Violence is rendered invisible, hierarchy naturalised, and oppression made ordinary (see Donoso, 2022).

What sustains the *scala naturae*, then, is not the force of argument or empirical observation but its entrenchment within social and political life. As Nussbaum notes, it rests on a belief system inhabited as a framework for living rather than examined as a philosophical position (Nussbaum, 2023, p. 23). Speciesist privilege stabilises this framework by ensuring that the benefits humans derive from animals appear justified while the harms on which those benefits depend remain unseen. In this way, subordination becomes ordinary, hierarchy inevitable, and animal oppression politically and morally unremarkable.

3 Care in a Multispecies Polity

3.1 From Privilege to Attentiveness

The analysis developed thus far has shown that oppression is sustained not only through social hierarchies and institutional arrangements but also through privileged dispositions that render it ordinary and morally unremarkable. Speciesist privilege sustains a hierarchical worldview in which the interests, vulnerabilities, and claims of animals are systematically ignored or rendered invisible. Against this background, the turn to care is a direct response to the political and epistemic failures identified in the previous section. If privilege works by normalising inattentiveness and insulating beneficiaries from responsibility, then the normative counterpoint must begin with a practice oriented toward attentiveness and responsiveness.

This is why hierarchy and care stand in direct tension. As Tronto observes, hierarchies "pose a threat to care" (Tronto, 2013, p. 164) – they undermine the very conditions for attentive, responsive, and responsible relations that care requires. In the context of human-animal relations, this tension is acute: the same hierarchical thinking that legitimises animal oppression actively forecloses the cross-species attentiveness that care would demand. If oppression is sustained by forms of distancing, disregard, and moral insulation, then care emerges as a necessary corrective.

Yet this immediately raises a difficulty. Care is often treated as a practice cultivated within close and intimate relationships, while public life is assumed to be governed by principles of justice. In this division, care appears ill-equipped to address structural injustices such as oppression, which operate at a distance, through institutions, and across entire social groups. If care were necessarily confined to intimacy and proximity, it would indeed seem incapable of responding to the forms of injustice analysed in this book. The question, therefore, is not whether care matters but whether it can be reconstructed in a form capable of responding to oppression as a structural injustice.

This worry has been articulated forcefully within the care ethics literature itself. As Stephanie Collins notes, caring deliberation is often characterised by sympathetic attention that involves "giving full attention to the person, while attempting to see the world as they see it from their perspective" (Collins, 2018, p. 195). Such accounts appear to presuppose conditions of closeness and familiarity largely absent from political institutions. Similarly, Alison Jaggar has warned that when care ethics remains focused on interpersonal relations, it risks losing critical purchase on social structures and therefore its capacity to function as a theory of justice (Jaggar, 1995, pp. 197–198).

Taken at face value, these critiques suggest a deep tension between care and the analysis of oppression as a structural injustice. But this tension does not arise from care as such. It arises from a restricted understanding of care that leaves intact the very hierarchies, distances, and epistemic blind spots through which oppression is reproduced. The following sections show how care can be reconstructed to meet this challenge.

Before proceeding, two clarifications are necessary. First, the analysis that follows focuses on the role of care in shaping political institutions rather than on particular policies. Institutions structure social life in deep and enduring ways, shaping not only individuals' positions within a political order but also the norms, expectations, and forms of agency available to them. Institutions are therefore a primary site at which structural injustices such as oppression are reproduced – and a primary site at which they must be confronted. Legal and political institutions are especially salient, since the principles governing social cooperation are made public, authoritative, and enforceable through them. Care-based policies are crucial (Hankivsky, 2004; Williams, 2001), but they are always designed, implemented, and constrained within broader institutional frameworks. A politics of care that fails to address those frameworks risks treating structural injustice as a series of isolated policy failures rather than as a systemic problem.

Second, the account developed here operates at a relatively high level of abstraction. Following a distinction drawn by Marcus G. Singer, I focus on institutions in the abstract sense – such as the state, the law, or the constitution – rather than on their particular concrete embodiments (Singer, 2003). This level of abstraction allows the identification of normative principles capable of guiding institutional design across different political contexts. More concrete illustrations and policy applications are taken up in later chapters. The aim here, by contrast, is to articulate a general normative framework for caring political institutions in a multispecies polity.

3.2 Can Institutions Care?

Let us reformulate the question guiding this part of the analysis: can institutions genuinely care? Can they be attentive and responsive to the needs, vulnerabilities, and dependencies of others? The main worry is that institutions do not provide intimacy, emotional presence, or personal attention, and therefore cannot replicate the kind of care characteristic of interpersonal care-giving (Noddings, 2015).

This objection fails, however, because it confuses an ethics of care with a politics of care. The ethics of care, as developed by Noddings and others, theorises care as an interpersonal relation between particular individuals, characterised by emotional presence and direct attentiveness. A politics of care has different requirements: it asks not that institutions feel for those they govern, but that they organise the social, legal, and material conditions in which attentiveness, responsibility, and responsiveness become structurally possible. Political institutions need not care as individuals do to be caring.

To see how institutions can be caring in their own right, it is useful to distinguish three of the four phases of care identified by Fisher and Tronto: caring-about, caring-for, and care-giving. The first phase concerns attentiveness to the needs of others. This phase is especially salient in institutional contexts, since oppression is often sustained precisely through failures of recognition and visibility. Political institutions can embody caring-about by making the recognition of others' needs and capacity for self-development a structural feature of their operation. Constitutions, for example, may enshrine principles that acknowledge all beings under institutional authority as vulnerable subjects whose needs warrant consideration (see Chapter 7). Likewise, administrative agencies can institutionalise attentiveness through procedures that ensure claims, complaints, and petitions are not merely received but taken seriously. Institutions can operate at the level of caring-about when they recognise needs and make them visible within public norms and procedures. Failures at this stage reproduce ignorance and invisibility – central mechanisms of oppression.

In contrast to caring-about, the next two phases – caring-for and care-giving – are more problematic. Caring-for involves taking responsibility for ensuring that needs are met, even when one is not the direct provider of care, while care-giving concerns the concrete organisation and delivery of care itself. It is often objected that political institutions and large organisations cannot meaningfully engage in these phases. Nel Noddings, for example, argues that institutions are necessarily limited to caring-about (Noddings, 2015, pp. 78–80) because they lack the intimacy and embodied responsiveness characteristic of interpersonal care.

But this rests on an unduly narrow understanding of responsibility. Institutions can assume responsibility for meeting needs through healthcare systems, social security, education, disaster response, and environmental protection. Kathleen Lynch's discussion of Monica, a 70-year-old widow, helps clarify what it means to be responsible for the care of someone without continuous physical presence.

Lynch describes how Monica sleeps with her mobile phone under her pillow, knowing that her daughter, who lives far away, does the same, ready to respond in case of an emergency. This arrangement, Lynch argues, mattered not because it involved constant interaction but because it established a standing assurance of availability: the possibility of a reliable response, rather than permanent presence or immediate intervention, can constitute a meaningful form of care (Lynch, 2022, p. 69).

Institutions can provide analogous assurances through publicly organised commitments that counter abandonment, neglect, and insecurity. Arrangements such as unemployment benefits, income guarantees, or universal health-care systems exemplify this form of institutional responsibility: they do not depend on intimacy but on public commitments to respond to vulnerability. Likewise, regulatory frameworks that ensure accessibility for people with disabilities, or institutional practices such as providing sign-language interpretation of official communications, constitute genuine forms of care-giving. These arrangements counter structural forms of neglect and exclusion, making social participation and self-development possible where oppression would otherwise foreclose them.

In sum, caring institutions are structured around recognition, reliability, and responsibility – conditions necessary for self-development and for embodying care in a meaningful, non-interpersonal sense.

3.3 The Politics of Care

Where oppression is structural, the response must be structural as well. If institutions can genuinely embody care, then care becomes a principle capable of operating at precisely the level where oppression is produced and sustained. I take this to be the core insight of what Joan Tronto and others describe as a politics of care (Tronto, 1993, 2013; Engster, 2007; Williams, 2001). On this view, care is not confined to the private sphere but shapes how institutions are organised, which needs are recognised as legitimate, and whose voices are included in processes of collective decision-making.

Tronto captures this when she writes that "care can serve as both a political concept to prescribe an ideal for a more democratic, more pluralistic politics [...] [and] as a strategic concept to involve the relatively disenfranchised in the political world" (Tronto, 1993, p. 21). Care here is not a sentiment or virtue but a normatively demanding political principle aimed at countering hierarchy, exclusion, and irresponsibility. Where privileged irresponsibility is sustained by distance and denial, a politics of care seeks to institutionalise recognition, responsibility, and responsiveness.

Although Tronto herself only gestures briefly toward human–animal relations, the political conception of care she develops provides the resources needed to address animal oppression as a structural injustice.[4] Care, however, must still be specified in a way that avoids reproducing the anthropocentric assumptions that have helped to sustain that oppression.

3.4 Toward a Non-Speciesist Conception of Care

Care must be stated at a level of generality that matches the scope of the political argument. In the context of structural oppression, care cannot be understood merely as a personal disposition or interpersonal virtue. At its most general level, care names a set of practices oriented toward attentiveness and responsiveness to others' needs, vulnerabilities, and dependencies – practices that sustain the conditions of self-development rather than undermine them, whoever the bearers of those needs may be.

A canonical articulation of this broader conception is offered by Berenice Fisher and Joan Tronto, who define care as:

> a species activity that includes everything that we do to maintain, continue, and repair our 'world' so that we can live in it as well as possible. That world includes our bodies, our selves, and our environment, all of which we seek to interweave in a complex, life-sustaining web.
>
> *(Fisher & Tronto, 1990, p. 40; Tronto, 1993, p. 103)*

This definition already gestures beyond narrow interpersonal models and explicitly includes the environment. Yet the "we" anchoring this account remains implicitly human. Caring practices are defined as those required to maintain the world so that humans can live in it as well as possible. Animals, if considered as part of the environment, appear not as subjects of care but as background conditions whose moral relevance is derivative of human self-development.

This anthropocentric framing is not incidental. It reproduces the very hierarchical logic that sustains speciesist oppression. One might attempt to rescue the account by arguing that caring-for animals is necessary for human self-development, but such a justification would remain fundamentally instrumental. Animals would still matter only contingently, as means to human ends.

An analogy clarifies the problem. Suppose someone claimed that caring practices exist to maintain and repair the world so that able-bodied people can live well. We would immediately object: what, then, of people with disabilities? Are they not equally entitled to care? Now imagine the defenders of this view responded that caring-for people with disabilities is, in fact, necessary for the development of the able-bodied, and therefore fully justified on those grounds.

Such a response would miss the point entirely. People with disabilities are owed care not because doing so benefits the able-bodied but because they are members of the political community with equal moral standing. Grounding their entitlement to care in the interests of others would reproduce precisely the kind of hierarchical reasoning that underpins their marginalisation. The same logic applies to nonhuman animals. If care for animals is justified only insofar as it enhances human well-being, then animals remain positioned as means rather than as subjects of justice.

Responding adequately to animal oppression, therefore, requires rejecting instrumental justifications and extending the "we" of care beyond the human.

Building on Fisher and Tronto, I propose the following reformulation: care is a species activity that includes everything that human and nonhuman animals do to maintain, continue, and repair a more-than-human world so that they can live and develop within it. That world includes their bodies, their selves, and their shared environment, all of which are interwoven in a complex, life-sustaining web.

Once animals are recognised as subjects of care, institutions cannot be designed solely around human interests. They must be responsive to the needs, vulnerabilities, and dependencies of all members of a multispecies polity.

3.5 Animals and Institutional Care

So understood, the implications of care for existing political institutions are stark. Currently, political institutions not only neglect animals; they often actively participate in arrangements that undermine their basic needs and capacities for self-development. An interspecies politics of care aims to transform these institutions into spheres of animal justice: institutional configurations in which political authorities assume responsibility for caring-for nonhuman animals as members of a shared multispecies polity.

To illustrate, consider the scale of institutional neglect in human–avian relations revealed by the phenomenon of bird–window collisions. In 2004, the British Trust for Ornithology estimated that approximately 100 million birds collide with windows each year in the United Kingdom, with severe consequences for already vulnerable species (BBC News, 2004). More recent research suggests that in the United States alone, up to one billion birds die annually from such collisions (Loss et al., 2014), making window strikes one of the leading anthropogenic causes of avian population decline in North America (Rosenberg et al., 2019). These figures do not reflect isolated accidents but reveal a systemic failure of institutional design: in the absence of regulation, the built environment itself becomes a structural hazard that predictably and repeatedly undermines birds' survival and capacity to develop their lives.[5]

This is not an inevitable feature of urban life. Political institutions possess both the authority and the capacity to transform these conditions. In 2020, the New York City Council enacted legislation requiring bird-safe materials in new and substantially altered buildings. Following renovations at the Jacob K. Javits Convention Center, bird–window collisions declined by more than 90 per cent (Pacheco, 2019). Beyond regulating private construction, political institutions can assume direct caring responsibilities by redesigning and maintaining public buildings in ways that mitigate avian fatalities and support birds' capacities for self-development – through bird-safe architecture, accessible water sources, and shelter in densely built or polluted urban areas.

This shows how institutions can move beyond abstract recognition to concrete caring-for. They redesign shared spaces so that animals can move, forage, and live without predictable harm. In doing so, they help constitute spheres of animal justice rather than spheres of oppression. Care thus emerges not as benevolence but as a principle of institutional design. When political institutions assume this role, they become active participants in shaping multispecies worlds – contexts in which nonhuman animals' needs and vulnerabilities are acknowledged and protected within the basic structure of society.

4 From Care to Solidarity

The analysis developed thus far has shown that institutions can embody care across their different phases and that such care is indispensable for countering structural oppression. Yet care, even when institutionalised, is insufficient on its own for the construction of animal justice in a full sense.

Care is oriented toward need, vulnerability, and dependency. It attends to what others lack and assumes responsibility for meeting it. Although this responsiveness is precisely what makes care indispensable in the face of oppression, it also marks its limit.

Care, even institutionalised care, can be extended paternalistically, as something done for others rather than with them as equals. A caring institution might attend fully to the needs of others, assume responsibility for meeting those needs, and provide material assurance against harm, while still positioning those being cared-for as objects of benevolent management rather than as subjects of justice. The history of welfare provision for marginalised human groups illustrates this pattern: care extended without equality, responsiveness without recognition of standing, benevolence that mitigates suffering while leaving intact the hierarchies that produce it.

Solidarity closes this gap – but in a specific way. Avery Kolers defines solidarity as acting on behalf of someone else, "acting as they would have acted if they could be in two places at once" (Kolers, 2018, p. 2). Solidarity is political action on others' terms (Kolers, 2016, p. 50). The emphasis on *others' terms* is decisive. What distinguishes solidarity from benevolence or well-intentioned care is the refusal to define the needs of the oppressed from the outside. Kolers is clear on this point: the belief that one's individual judgement can and should supersede the collective terms set by those who struggle is not help but imposition – an erosion of their efficacy that functions, even when well-meaning, as what Kolers describes as an "attack" (Kolers, 2018, p. 4). To act in solidarity is not to intervene as a saviour but to place one's capacities at the disposal of the struggle, on its terms (Kolers, 2016, 2018).

Applied to animals, this means that solidarity requires a genuine effort to respond to their needs and vulnerabilities as expressed in their own forms of life – in their behaviours, their developmental trajectories, their species-specific modes of flourishing – rather than in terms that happen to be convenient or legible to human institutions.

Kolers also argues that solidarity is owed as a duty, not as a moral option – as a debt of equitable treatment owed to particular beings at particular times. Failing to respond to the call for solidarity does not leave one neutral: it means continuing as though the treatment of the oppressed were appropriate, treating them as one more object rather than as an equal (Kolers, 2018, p. 3). This has uncomfortable precision in the context of animals, who are routinely classified – legally, economically, institutionally – as exactly that: objects. To extend solidarity to animals is to rebuke that classification in practice, not merely to deplore it in principle. It is to refuse to go on benefiting from arrangements that treat them as disposable, and to share, as far as one's situation allows, in the fate to which those arrangements consign them.

There are, however, two related difficulties. Kolers grounds the duty of solidarity in equity and ties its exercise to the capacity of the oppressed to issue a call – to name the site of their struggle and demand that others act alongside them. Animals cannot issue such a call in Kolers's sense. Also, Kolers's account is, throughout, an account of solidarity as an individual virtue: what particular persons owe to other particular persons when they witness inequitable treatment. The argument developed here operates at a different level. Its primary concern is not with what individuals owe each other but with what political institutions owe to all those whose lives they shape and regulate.

These difficulties do not dissolve the obligation, but reframe it. The account of solidarity developed in this chapter extends Kolers's logic in two directions. First, it relocates the duty of solidarity from individuals to institutions: it is states, legal systems, and public arrangements that bear the primary obligation to treat all inhabitants equitably, and that fail in that obligation when they permit, normalise, or actively reproduce the oppression of animals. Second, it regrounds the duty not in the call – which presupposes capacities for political agency that animals lack – but in the fact of shared inhabitance: animals are subject to the same institutional arrangements that structure human life, and that subjection is itself sufficient to ground the demand for equitable treatment. Solidarity with animals, institutionally understood, is owed not as charity but as equity – as the recognition that their interests carry equal weight independently of their capacity to demand it.

The combination of care and solidarity, understood in these institutional terms, does more than patch a deficit in either principle taken alone. It redefines what just institutions in a multispecies society must look like: not merely responsive to vulnerability, but accountable to equality; not merely attentive to need, but committed to meeting it on the terms of those whose needs they are. Together, care and solidarity provide the normative foundations for transforming institutions from sites of oppression into genuine agents of interspecies justice.

4.1 Solidarity, Care, and Oppression

Unlike care, solidarity targets the political relations through which vulnerability is distributed and rendered unequal. Solidarity affirms the equal standing of all inhabitants of a political community, resists marginalisation, and transforms care from a merely benevolent response into a demand grounded in justice. It provides the explicitly political orientation that care, on its own, cannot secure. Whereas care foregrounds responsiveness to vulnerability, solidarity directs attention to the social relations through which vulnerability is produced, distributed, and entrenched.

Despite these differences, care and solidarity share important structural features. Both are irreducibly relational: one does not simply care, but cares about or for others; solidarity likewise is necessarily oriented toward the struggles and claims of others rather than exercised in isolation. Both resist atomistic accounts of social life, emphasising relations of interdependence and mutual implication. Both involve positive obligations. Care is not exhausted by refraining from harm but requires attending to needs and, where necessary, providing support. Solidarity, similarly, cannot be reduced to non-interference; it demands action aimed at supporting others and contesting the conditions under which they are marginalised.

At the level of individual political practice, solidarity means aligning with those subjected to oppression and committing to challenge the hierarchies that constrain their self-development. At the institutional level – which is the primary concern here – it means structuring political arrangements so that equal standing is not contingent on the goodwill of those in positions of care, but is secured as a requirement of justice. These two levels are complementary rather than competing: individual solidarity without institutional backing remains fragile and episodic, while institutions structured by solidarity without the animating commitment of those who inhabit and operate them risk becoming hollow forms. What an interspecies politics of care requires is both – and it requires that the institutional level be understood as primary, since it is institutions that determine, in the most durable and far-reaching way, whether the equal standing of animals is treated as a matter of justice or left to the discretion of the benevolent.

Joan Tronto acknowledges this political dimension by identifying solidarity as central to a fifth phase of care – caring-with – in which the meeting of needs is integrated with democratic commitments to justice, equality, and freedom (Tronto, 2013, pp. 23, 35). This phase captures the insight that care becomes fully politically transformative when exercised in ways consistent with equal standing and collective self-rule. In this respect, Tronto's account converges with Sally J. Scholz's conception of social solidarity, which emphasises the interdependence binding inhabitants of a political community (Scholz, 2008). Solidarity here is not an affective disposition but a relational and institutional practice through which mutual responsibility and common inhabitance are made politically operative.[6]

Solidarity is not grounded merely in shared sentiment or identity but in the constitutive dependencies that bind members of a community together and demand collective responsibility. Once extended beyond the human, those constitutive dependencies encompass the interdependencies that tie humans and other animals together within shared political and institutional arrangements. Once vulnerability and dependency are understood not as exceptional conditions but as permanent features of social and political life, solidarity becomes indispensable for guiding how care is organised and distributed.

This raises a further question: what form does solidarity take when injustice persists despite institutional arrangements? Alasdair Cochrane and Mara-Daria Cojocaru (2023) move in this direction by framing veganism as a form of political solidarity. Building on Scholz (2008), they stress veganism's explicitly political character: its collective resistance to structures of oppression, particularly those embedded in industrial animal agriculture, biomedical research, and the entertainment industry (Cochrane & Cojocaru, 2023, p. 62). What matters, on their account, is not whether individuals have a duty to perform some specific act, but whether they should join with others to pursue collective goals such as confronting injustice. Veganism thus illustrates how solidarity becomes politically salient when injustice is identified and collectively resisted.

For this chapter, however, the significance of political solidarity extends beyond the practices of particular social groups. While Cochrane and Cojocaru rightly connect veganism with social justice and emphasise solidarity as a mode of resistance to oppression, their analysis remains focused on collective action among individuals. My concern here is different: to highlight the role of solidarity at the level of political institutions within an interspecies politics of care. If solidarity is indispensable for confronting injustice, it cannot remain confined to citizen action alone. The following section develops that institutional dimension by turning to a further form of solidarity – one grounded not in the practices of social groups but in the obligations of political communities toward all those whose lives they shape.

4.2 Civic Solidarity

Sally J. Scholz identifies a form of solidarity that is especially relevant for the institutional focus of my argument: civic solidarity. This form concerns the obligations that a political community, acting through the state, has toward those who fall under its authority. Scholz defines civic solidarity as referring to "the obligations the state as a collective has to each citizen" and to the role of public institutions in protecting individuals against vulnerabilities that would otherwise prevent their participation in social and political life (Scholz, 2008, p. 27) – a formulation this chapter extends beyond citizens to all inhabitants of a political community. At its core, civic solidarity names the responsibility of political institutions to secure the basic conditions under which members of a community can live, act, and develop without being systematically exposed to neglect or oppression.

Once this idea is reframed in a non-anthropocentric register, its implications become clearer. If civic solidarity grounds institutional obligations in contexts structured by political authority, its scope cannot be limited to humans alone.

Importantly, this does not require assuming that political authority is exercised over nonhuman animals. Rather, political authority is exercised over human political agents and institutions, which are thereby obligated to act in particular ways toward nonhuman animals. Through law, regulation, and public policy, states structure the conditions under which animals live, are used, protected, or harmed, and in doing so profoundly shape their vulnerabilities and life prospects.

Civic solidarity therefore requires that political institutions recognise nonhuman animals as subjects of care in their own right. This recognition is realised centrally – though not exclusively – through institutionalised practices of care: sustained attentiveness to needs and vulnerabilities, and the assumption of responsibility for meeting them. In this view, animals are not citizens or members of the political community in the conventional sense – they are not bearers of political authority, nor participants in collective self-rule. But they are unavoidably subject to political decision-making, and it is precisely this fact – their status as inhabitants of a shared political space whose lives are shaped by institutional arrangements – that grounds institutional obligations toward them.

Civic solidarity also affirms equal political standing within a community of justice. By political standing, I do not mean the capacity to participate as political agents – a capacity that grounds citizenship – but the requirement that interests be taken seriously and given equal consideration in the organisation of social and political institutions. This standing is grounded not in membership but in inhabitance: in the fact of sharing a world whose political and institutional arrangements determine the conditions under which self-development is possible or foreclosed. What matters is the depth of this inhabitance: institutions that structure the most basic conditions of a being's existence – where it lives, what it consumes, whether it can reproduce, whether it lives or dies – cannot be just unless they take that being's interests into equal account. Understood in this way, an interspecies politics of care demands that states and public institutions extend equal consideration to all inhabitants of the political order, human and nonhuman alike. Civic solidarity thus marks the point at which care ceases to be discretionary and becomes a requirement of justice.

Civic solidarity and political solidarity stand in a dynamic and complementary relation. Civic solidarity specifies the institutional conditions required to secure equal political consideration for all those whose lives are shaped by state authority – citizens and non-citizens, humans and nonhuman animals alike. Political solidarity, by contrast, mobilises individuals and groups to resist injustice in contexts where institutions fail to realise the demands of civic solidarity. Political solidarity therefore becomes especially salient when institutional obligations are unmet or systematically violated. Where civic solidarity is robust and effective, the oppositional function of political solidarity is correspondingly reduced – though never entirely eliminated.

At the opening of this section, I noted that both care and solidarity impose positive demands on agents. If care requires responsiveness to the needs of others, solidarity obliges agents to join with others in struggles for equality. It is precisely this commitment to equality that grounds the obligations civic solidarity imposes on political institutions. Civic solidarity can therefore be understood as an institutional form of resistance to hierarchical thinking and practice, not merely a guiding ideal but a way of performing equality within political life.

Institutions act with civic solidarity when they respond adequately to the needs and vulnerabilities of all those whose lives they shape and regulate, thereby affirming equal standing in practice rather than in name alone. By contrast, institutions that care for some but not others, or that selectively support the self-development of members of certain groups only, act without solidarity. Such institutions cannot be just and have no place in a genuinely just political community.

This understanding of civic solidarity complicates the role of political institutions in multispecies communities. Meeting its demands requires sustained attentiveness to the plurality of forms of life and self-development that coexist within the political order, as well as ongoing revision of institutional practices to ensure that care is extended inclusively rather than restricted by anthropocentric or other hierarchical assumptions. Civic solidarity fails whenever institutions reproduce relations of inequality – whether through racism, sexism, ableism, or speciesism. In such cases, political solidarity re-emerges as a necessary form of resistance against the oppression enabled or enacted by institutional arrangements.

A just multispecies political community must therefore be grounded in civic solidarity understood as an institutional commitment to radical equality. Its radicalism lies both in rejecting anthropocentrism and in affirming the obligation of political institutions to support the self-development of all those whose lives they shape. This commitment defines the interspecies politics of care advanced in this chapter.

5 Conclusion

I have presented the contours of an interspecies politics of care applicable to political institutions within a non-anthropocentric framework. By rearticulating Fisher and Tronto's account of care in a non-speciesist register, I have argued that political institutions can, and ought to, provide care to those whose lives they shape. Contrary to sceptical views that confine care to the interpersonal domain, political institutions can act as caring institutions.

At the same time, care alone is insufficient to render political practice responsive to the demands of justice in the face of oppression. Care's normative foundation is responsiveness to need, and it cannot by itself ground equal standing – a standing owed to beings not because they are currently vulnerable but because they share a common political order. For this reason, I have argued that a politics of care must be complemented by solidarity if it is to acquire full political force. Drawing on and extending Scholz's concept of civic solidarity, I have shown how political institutions can be shaped by obligations grounded in equality – obligations to respond to the needs and vulnerabilities of others as equals, and to resist the hierarchies that render some lives disposable. Within a non-hierarchical framework, civic solidarity expands the scope of justice to include all those whose lives are structured by institutional arrangements, humans and nonhuman animals alike. This, I suggest, marks the normative horizon of institutions in a just multispecies society.

Chapters 7, 8, and 9 examine three institutional domains central to the human–animal interface in industrialised societies: constitutional norms, urban organisation, and food production systems. Each of these domains currently reproduces hierarchies and privilege, transforming social, economic, and political spaces into spheres of oppression. The chapters that follow explore how these institutions could be reconfigured as spheres of animal justice – spaces structured by care and solidarity, capable of resisting oppression and enabling self-development across species.

Notes

1 Particularly, Tronto examines how privileged white women in US higher education often fail to recognise the structural and interpersonal violence experienced by Black women, illustrating how racial privilege can coexist with gender-based disadvantage.
2 The post-1990 culture of disavowal among former regime supporters in Chile has been analysed extensively by Stern (2010). See also Stern 2004 and 2006.
3 The invisibilisation of animal suffering in these industries is reinforced by well-documented psychological mechanisms. Consumers routinely justify meat consumption through what researchers call the "4Ns" – the beliefs that eating meat is natural, normal, necessary, and nice – which function to reduce moral discomfort and narrow the circle of beings deemed worthy of ethical concern (Piazza et al., 2015). Where dissonance persists, it is managed through denial of animals' mental capacities and, in some cases, through strategic ignorance: the active avoidance of information about animal sentience in order to protect existing habits from moral challenge (Rothgerber & Rosenfeld, 2021). I am grateful to Frederike Spang for directing me to this research.
4 For example, she praises democracy for being the best political regime because it permits better than alternative forms of government "to care for one another, *for other animals* and things in the world, and for the world itself" (Tronto, 2013, p. 156. My emphasis).
5 The concern for the deaths of birds due to collisions with human-made buildings is not new. In the 19th century, the British ornithologist William Eagle Clarke documented the increasing number of migratory birds dying after crashing into the lanterns of light stations on the coasts of Ireland and Great Britain, particularly in foggy weather (Eagle Clarke, 1896, pp. 622, 626).
6 On the grounding of solidarity in constitutive interdependencies rather than shared sentiment or identity, see Jaeggi (2001, p. 297). Once extended beyond the human, that relational fabric encompasses the interdependencies tying humans and other animals together within shared political and institutional arrangements – a move that further supports the non-anthropocentric extension of civic solidarity developed in this chapter.

References

BBC News. (2004, March 11). Birds killed by window strikes. http://news.bbc.co.uk/2/hi/science/nature/3505256.stm.

Bowman, M. (2020). Privileged ignorance, "world"-traveling, and epistemic tourism. *Hypatia*, 35(3), 475–489. doi:10.1017/hyp.2020.26.

Bozalek, V. (2014). Privileged irresponsibility. In G. Olthuis, H. Kohlen, & J. C. Tronto (Eds.), *Moral boundaries redrawn: The significance of Joan Tronto's argument for political theory, professional ethics, and care as practice* (pp. 51–72). Peeters.

Ceryes, C. A., & Heaney, C. D. (2019). "Ag-gag" laws: Evolution, resurgence, and public health implications. *New Solutions*, 28(4), 664–682.

Cochrane, A., & Cojocaru, M.-D. (2023). Veganism as political solidarity: Beyond 'ethical veganism'. *Journal of Social Philosophy*, 54(1), 59–76. doi:10.1111/josp.12460.

Collins, S. (2018). Care ethics: The four key claims. In D. Morrow (Ed.), *Moral reasoning: A text and reader on ethics and contemporary moral issues* (pp. 192–204). Oxford University Press.

Donoso, A. (2022) Ecología Integral en Laudato Si'. Hacia una Educación Ciudadana Ecológica. In Rodrigo Mardones (Ed.), *Educación Ciudadana y Enseñanza Católica: La Fraternidad como Desiderátum*. Ediciones Universidad Católica de Chile.

Eagle Clarke, W. (1896). Bird migration in the British Isles. *Scottish Geographical Magazine*, 12 (12), 616–626.

Engster, D. (2007). *The heart of justice: Care ethics and political theory*. Oxford University Press.

Fisher, B., & Tronto, J. (1990). Towards a feminist theory of care. In E. Abel & M. Nelson (Eds.), *Circles of care* (pp. 35–62). State University of New York Press.

Hankivsky, O. (2004). *Social policy and the ethics of care*. University of British Columbia Press.

Jaeggi, R. (2001). Solidarity and indifference. In R. ter Meulen, W. Arts, & R. Muffels (Eds.), *Solidarity in health and social care in Europe* (pp. 287–308). Kluwer Academic Publishers.

Jaggar, A. M. (1995). Caring as a feminist practice of moral reason. In V. Held (Ed.), *Justice and care: Essential readings in feminist ethics* (pp. 179–202). Westview Press.

Loss, S. R. et al. (2014). Bird–building collisions in the United States: Estimates of annual mortality and species vulnerability. *The Condor*, 116(1), 8–23.

Kolers, A. (2016). *A Moral Theory of Solidarity*. Oxford University Press.

Kolers, A. (2018). *The Moral Duty of Solidarity*. Civil American: Philosophy for Everyone 3: 1–6.

Lynch, K. (2022). *Care and capitalism*. Polity Press.

McCausland, C. et al. (2013). Trespass, animals and the democratic process. *Res Publica*, 19, 205–221.

Medina, J. (2013). *The epistemology of resistance: Gender and racial oppression, epistemic injustice, and the social imagination*. Oxford University Press.

Noddings, N. (2015). Care ethics and "caring" organizations. In D. Engster & M. Hamington (Eds.), *Care ethics and political theory* (pp. 72–91). Oxford University Press.

Nussbaum, M. C. (2023). *Justice for animals: Our collective responsibility*. Simon & Schuster.

Pacheco, A. (2019, September 9). Architects lobby for bird-friendly glass as NYC overhauls cladding regulations. *Archinect News*. https://archinect.com/news/article/150157872/architects-lobby-for-bird-friendly-glass-as-nyc-overhauls-cladding-regulations.

Pachirat, T. (2011). *Every twelve seconds: Industrialized slaughter and the politics of sight*. Yale University Press.

Piazza, J. et al. (2015). Rationalizing meat consumption: The 4Ns. *Appetite*, 91, 114–128.

Plumwood, V. (1993). *Feminism and the mastery of nature*. Routledge.

Robbins, J. A. et al. (2016). Awareness of ag-gag laws erodes trust in farmers and increases support for animal welfare regulations. *Food Policy*, 61, 121–125.

Rosenberg, K. V. et al. (2019). Decline of the North American avifauna. *Science*, 366 (6461), 120–124.

Rothgerber, H., & Rosenfeld, D. L. (2021). Meat-related cognitive dissonance: The social psychology of eating animals. *Social and Personality Psychology Compass*, 15 (5), e12592.

Scholz, S. J. (2008). *Political solidarity*. Pennsylvania State University Press.

Singer, M. G. (2003). *The ideal of a rational morality: Philosophical compositions*. Oxford University Press.

Stern, S. J. (2004). *Remembering Pinochet's Chile: On the eve of London 1998*. Duke University Press.

Stern, S. J. (2006). *Battling for hearts and minds: Memory struggles in Pinochet's Chile, 1973–1988*. Duke University Press.

Stern, S. J. (2010). *Reckoning with Pinochet: The memory question in democratic Chile, 1989–2006*. Duke University Press.

Tronto, J. C. (1990). Chilly racists [Paper presentation]. Annual Meeting of the American Political Science Association.

Tronto, J. C. (1993). *Moral boundaries: A political argument for an ethic of care*. Routledge.

Tronto, J. C. (2013). *Caring democracy: Markets, equality, and justice*. New York University Press.

Van den Brandeler, E. (2024). Towards an epistemology of 'speciesist ignorance'. *Res Publica*, 30(4), 783–804.

Williams, F. (2001). In and beyond New Labour: Towards a new political ethics of care. *Critical Social Policy*, 21(4), 467–493.

Wilson, L. (2014). Ag-gag laws: A shift in the wrong direction for animal welfare on farms. *Golden Gate University Law Review*, 44(3), 311–335.

7

CONSTITUTIONALISING ANIMAL JUSTICE

1 Introduction

Following the previous chapter's defence of care and solidarity as normative responses to structural injustice in a multispecies polity, this and the following two chapters turn to institutions. If oppression is embedded in the basic structure of society, dismantling it requires more than ethical orientation; it demands institutional transformation. The task is to ask how the normative framework of an interspecies politics of care can be translated into concrete political forms.

This chapter focuses on one foundational institution: the constitutional framework. It asks how a constitution might recognise animals as subjects of justice without reducing them either to passive objects of protection or to resources subordinated to human welfare. Building on emerging legal innovations – from the constitutional recognition of the rights of nature in Ecuador and Bolivia to Germany's constitutional commitment to animal protection – it explores how constitutional design can expand moral and political consideration beyond the human to incorporate animals as a matter of justice.

To illuminate this possibility, the chapter introduces the concept of a caring constitution: a legal architecture that embeds interspecies responsibility into the very definition of justice and into the organisation of the state. A caring constitution does not merely add animal-protection clauses to an otherwise anthropocentric framework. Rather, it reshapes public institutions, policy-making procedures, and forms of civic participation so that environmental, agricultural, economic, and urban governance must integrate nonhuman interests as a matter of justice.

The caring constitution thus represents the institutionalisation of care and solidarity as the normative core of a multispecies polity. The rest of this

DOI: 10.4324/9781003434856-9

chapter explores what this looks like in practice: through existing constitutional experiments, through three institutional domains where the state's authority is already constitutionally engaged, and through the civic ethos required to sustain such commitments over time.

2 The Constitution of the Multispecies State

As shown in the first part of this book, human–animal relations are structured – implicitly and explicitly – through state practices across multiple domains: agriculture, environmental regulation, urban planning, trade, and public health. After that diagnosis, the question shifts. Rather than asking how the state in fact organises human–animal relations, I ask how it ought to do so if it is to align its basic structure with justice.

Because constitutions occupy a foundational legal and symbolic position within institutional orders, they are a privileged site for articulating the principles that legitimise state power within a multispecies political community. This is not by itself decisive in favour of constitutional inclusion. It does, however, generate a strong presumptive case. If animals are genuine subjects of justice – as Part I argued – those who would exclude them from constitutional protection bear the burden of justifying that exclusion.

Constitutional norms do not merely regulate policy; they define the horizon within which law and policy become intelligible. Embedding principles of interspecies care and solidarity at the constitutional level therefore represents a transformative intervention. It orients the interpretation of laws, constrains legislative and administrative action, and signals a commitment to reorganising human–animal relations as a matter of justice.

Constitutional frameworks are not conceptually uniform. They may be grounded in diverse theories of sovereignty, constituent power, or political morality. Yet whatever one's broader account of constitutional authority, constitutions share core features. They function as normative frameworks that organise and limit state power by defining both the ends for which coercion may be exercised and the goods a polity commits itself to pursuing. They distribute opportunities, authorise institutions, and establish the conditions under which power is rendered legitimate within the territory. In this sense, constitutions do more than merely reflect social values; they actively shape the normative architecture of political life.

If this is correct, then the inclusion or exclusion of animals from constitutional recognition is not a neutral organisational choice. It is a foundational political statement about who belongs, and how, to the community of justice. Constitutional silence is not neutrality; it amounts to an endorsement of an anthropocentric boundary of justice.

In contrast, interspecies justice favours the constitutional recognition of animals as subjects of justice. Such recognition rests on acknowledging animals as self-developing beings – agents whose lives unfold along their own developmental trajectories, marked by vulnerability, interdependence, social belonging, and their own ways of flourishing – and on recognising that this fact has implications for how state power is organised and constrained.

Admittedly, constitutionalisation is not the only route to interspecies justice. Alternative legal, cultural, or institutional pathways might, in principle, produce similar protective outcomes. Yet constitutional recognition carries a distinctive normative force. To recognise animals constitutionally is to inscribe care and solidarity into the foundations of collective life. It affirms that justice extends to all who possess the capacity for self-development under conditions of vulnerability and interdependence. This is the normative horizon a caring constitution must articulate.

3 Constitutional Footholds for Animal Justice

Although interspecies justice remains marginal within mainstream constitutional theory, a growing number of constitutional experiments already challenge the assumption that political communities are necessarily human-exclusive. These developments are fragmented and uneven. Most constitutional orders remain deeply anthropocentric, and existing provisions often operate indeterminately or sit uneasily alongside normative commitments that continue to authorise the oppression of nonhuman life. Yet their significance lies less in what they fully achieve than in what they render visible. They demonstrate that constitutions can serve as sites for rearticulating the principles of political morality that ground state authority and, in doing so, can reopen the question of who properly belongs within the political community.

These developments do not, in most cases, straightforwardly confer constitutional rights on animals. Rather, they create interpretive footholds – normative openings through which animals may be understood as more than objects of regulation or resources for human welfare. Through such openings, animals can be approached as subjects whose lives carry moral and political salience. Constitutional law, read in this way, appears not merely as a framework of restraint but as an institutional architecture capable of expressing non-anthropocentric care and solidarity through positive obligations imposed on public power.

The promise of these innovations lies in their capacity to reorient constitutional legitimacy toward the conditions that enable diverse forms of life to coexist and self-develop. In doing so, they challenge the image of human sovereignty over nature and gesture instead toward vulnerability and interdependence as foundational principles of political authority.

Against this background, I turn to a selection of these constitutional innovations – not as models to be replicated, but as resources for clarifying both the normative possibilities and the structural limits of a caring, non-anthropocentric constitutional order.

3.1 Latin America: Rights of Nature and Constitutional Ecology

Latin America offers some of the most explicit constitutional challenges to anthropocentric political ordering, particularly through the incorporation of ecological and animal-protective norms into constitutional texts. Ecuador's 2008 Constitution stands as the most far-reaching example. By explicitly recognising the rights of nature (*Pachamama*) and granting ecosystems legal standing to exist, regenerate, and be restored, Ecuador decisively disrupts the assumption that constitutional subjects must be human. Although animals are not expressly recognised as subjects of justice, the constitutionalisation of nature's rights destabilises the conventional subject–object distinction that underpins anthropocentric constitutionalism. It opens constitutional interpretation to nonhuman life as normatively significant in its own right, rather than merely as a resource to be managed for human ends.

This interpretive possibility was made explicit by the Constitutional Court of Ecuador in 2022 in the case of Estrellita, a chorongo monkey who had lived for years with a woman in Ambato (Montes Franceschini & Stilt, 2023). Interpreting Article 71 of the Constitution, the Court affirmed that legal actions may be brought in the name of Nature to demand the protection and reparation of its integrity, including its constituent animals.[1] By this declaration, the Court recognised that constitutional protection extends not only to collective ecological wholes but also to their constituent parts, including wild animals. From the perspective of interspecies justice, Ecuador's constitutional framework does not yet articulate animals as subjects of justice. Nevertheless, it renders such recognition conceptually intelligible by unsettling the anthropocentric boundary of constitutional concern.

Brazil's 1988 Constitution (Ch. VI, Art. 225) represents a more restrained, yet still consequential, approach. It mandates the protection of fauna and flora and requires the state to prevent practices that cause the extinction of species and subject animals to cruelty (Di Concetto, 2017). These provisions embed animal protection within a broader ecological duty of the state and impose constitutional constraints on both public and private action. However, they remain framed primarily in terms of welfare and environmental protection rather than justice. Animals appear as beings to be shielded from excessive suffering, not as cohabitants of a political community whose self-development imposes structural demands on institutional design. Even so, by constitutionalising anti-cruelty obligations, Brazil's Constitution acknowledges animals' vulnerability to systemic harm and limits the legitimacy of state practices that expose them to institutionalised violence.

In previous work (Donoso, 2021), I have argued that rights-of-nature frameworks frequently leave unresolved the relationship between collective ecological goods and the rights of individual animals. In situations of conflict, collective ecological goals may be invoked to justify harm to vulnerable individuals – and without a principled account of how collective and individual rights relate, the recognition of the rights of nature risks undermining the protective function that rights are meant to serve. From the standpoint of interspecies justice, a caring constitution cannot leave this tension unaddressed.

3.2 Europe: State Responsibility, Dignity, and Constitutional Limits

In Europe, constitutional engagement with animals has generally taken a more incremental and institutionally cautious form. Rather than conferring rights, European constitutions tend to frame animal protection in terms of state responsibility, environmental stewardship, or limits on excessive instrumentalisation.

A paradigmatic example is Germany's 2002 amendment to Article 20a of the Basic Law.[2] By inserting animals alongside the "natural foundations of life," the amendment constitutionally obliges the state to protect animals through legislative, executive, and judicial action. This reform elevated animal protection from a matter of legislative discretion to a constitutional objective binding on all branches of government. Although animals are not recognised as rights-holders, the amendment reframes them as constitutionally relevant beings whose treatment bears on the legitimacy of state action. German constitutional jurisprudence has treated Article 20a as a genuine normative constraint, capable of limiting policymaking even where economic or scientific interests are at stake. Still, the provision remains structured as a duty of the state, not as a recognition of animals as subjects of a justice-bearing community.

Switzerland has gone further by constitutionalising the dignity of living beings. Article 120(2) of the 1999 Constitution provides that, in regulating reproductive and genetic technologies, the Confederation shall take account of the dignity of living beings and protect biodiversity. Although introduced within the context of biotechnology, this clause has broader interpretive implications. By attributing dignity beyond the human, the Constitution unsettles the assumption that dignity is an exclusively anthropocentric concept. The notion remains contested and unevenly applied, yet its constitutional status requires that animals be treated as more than mere objects of use. From the standpoint of interspecies justice, Switzerland's model approaches a non-anthropocentric constitutional imagination more closely than most European systems. Nevertheless, dignity here functions primarily as a constraint on excessive instrumentalisation rather than as a foundation for positive claims to self-development.

More recently, Luxembourg's 2023 revision of its Constitution introduced an even clearer ontological shift. Article 41(3) now provides that the state recognises animals as nonhuman living beings endowed with sentience and ensures the protection of their welfare. This language does not merely impose a duty of protection; it affirms animals' status as sentient beings. Although the provision remains tied to welfare protection rather than justice claims, it signals a movement away from purely environmental framing toward recognition of animals as individual bearers of morally relevant characteristics.

Taken together, these European developments reveal a constitutional trajectory that moves from environmental duty (Germany), to dignity-based constraint (Switzerland), to sentience recognition (Luxembourg). Yet they stop short of articulating animals as self-developing subjects whose interests structure the basic design of political institutions. They constrain state action; they do not yet redefine the political community. From the perspective of a caring constitution, their significance lies precisely in this transitional character. They demonstrate that constitutional orders can acknowledge nonhuman moral status, but they leave unresolved the deeper question of how care and solidarity might be embedded as organising principles of state authority itself.

3.3 Asia: Judicial Interpretation and Constitutional Expansion

In parts of Asia, constitutional engagement with animals has emerged less through textual reform than through judicial interpretation. India provides the most significant example. Rather than amending the Constitution to recognise animal rights, the Supreme Court has interpreted existing provisions in ways that expand constitutional concern beyond the human.

The landmark case is *Animal Welfare Board of India v. A. Nagaraja* (2014). Interpreting Article 21 – the constitutional guarantee of the right to life – the Court affirmed that animals possess intrinsic worth and are entitled to live with dignity. Drawing on constitutional values such as life, compassion, and moral responsibility, the Court held that the protection of animals is not merely a matter of statutory welfare but of constitutional principle. Although the Court's judgment does not amount to a structural recognition of animals as subjects of justice, it demonstrates how constitutional values can be interpreted in ways that destabilise strict anthropocentrism and expand the moral scope of political authority.

The Court's reasoning also relied on Article 51A(g), which establishes the fundamental duty of every citizen to protect and improve the natural environment and to have compassion for living creatures. In Nagaraja, the Court interpreted this provision expansively, articulating a constitutional language of intrinsic worth and dignity that extends beyond the human. It reframed compassion not as a private virtue but as a constitutional value capable of shaping public authority.

Yet this jurisprudential expansion has proven fragile. Subsequent judicial and political developments in India have significantly narrowed the practical reach of Nagaraja, revealing the limits of a constitutional transformation driven primarily by interpretive activism rather than structural institutional redesign (Kansal, 2016). When constitutional commitments to animals are not embedded in the basic architecture of state authority, they remain vulnerable to reversal, dilution, or political override.

The Indian experience therefore illustrates both the promise and the instability of judicial constitutionalism. Courts can reopen the moral meaning of constitutional texts and momentarily expand the community of concern. But without a broader reorganisation of constitutional principles – one that inscribes care and solidarity into the foundations of legitimacy itself – such advances remain contingent and reversible.

3.4 From Constitutional Experiments to Normative Reconstruction

From the perspective of a caring constitution, the developments surveyed above do not amount to a settled model of just interspecies constitutionalism. None fully recognises animals as subjects of justice whose self-development must shape the basic design of political institutions. At most, they impose constraints on excessive harm, expand interpretive horizons, or introduce duties of protection. They do not yet reconfigure constitutional legitimacy around the idea that nonhuman beings form part of the community whose development and vulnerability structure the ends of state authority.

This distinction is decisive. To protect animals is not yet to recognise them as cohabitants of a political order bound by justice. A constitution grounded merely in protection retains the underlying asymmetry of anthropocentric authority: humans remain the sole authors and beneficiaries of political power, while animals appear as objects of regulation, compassion, or stewardship. What is absent is the more demanding claim that animals, as self-developing beings, impose principled limits on how coercive institutions are organised and justified.

Yet these constitutional experiments remain normatively significant. Their importance lies less in what they definitively settle than in what they render imaginable and institutionally available. Across continents, they demonstrate that constitutions are not fixed expressions of a human-exclusive political community but sites of contestation capable of incorporating nonhuman life into the domain of political concern. They unsettle the assumption that constitutional subjecthood must coincide with humanity and reveal that the boundaries of political community are historically contingent and institutionally constructed.

These developments are therefore best understood as constitutional footholds: partial, unstable, and often internally contradictory, yet nonetheless transformative. They show that constitutional law can be reoriented toward forms of care and solidarity that extend beyond the human. In doing so, they open space for rethinking legitimacy not as sovereignty over nature but as responsibility within a shared ecological community.

It is from these footholds that the next sections proceed. The aim is not merely to constitutionalise animal protection but to articulate a model in which animals are recognised as self-developing subjects of justice and in which interspecies care and solidarity are embedded within the very structure of political authority.

In what follows, I do not attempt to draft a comprehensive blueprint for a caring interspecies constitution. Rather, I focus on three domains in which constitutional authority has historically shaped the basic structure of social life: education, healthcare, and food systems.

The selection of these three domains is not arbitrary. Each satisfies two conditions that jointly distinguish them as primary sites of constitutional concern within a caring multispecies polity. First, each is a domain in which state authority is already constitutionally engaged in most political orders – education, healthcare, and food systems are universally recognised as matters of public responsibility, even where their precise constitutional status varies. Second, and more importantly for the present argument, each is a domain in which the structural oppression of animals is not incidental but constitutive: industrial food production is the primary institutional mechanism through which animals are commodified and killed at scale; healthcare governance determines whether animal health is treated as a matter of justice or merely of risk management; and education is the primary site through which the speciesist norms that sustain both are reproduced and naturalised across generations. The caring constitution's intervention in these three domains is not exhaustive of what interspecies justice requires – property law, criminal law, labour law, and urban planning all raise analogous constitutional questions, some of which are addressed in the following chapters – but it targets the institutional core of speciesist oppression as it operates within the basic structure of contemporary political life.

Food systems may appear to fall outside the traditional core of constitutions, but this impression rests on an unduly narrow understanding of constitutional scope. Constitutions routinely structure the legal architecture that governs agriculture, property, trade, public health, environmental protection, and increasingly the right to food itself. Because industrial food systems are among the primary institutional sites in which structural violence against animals is organised and normalised, they fall within any constitutional project concerned with interspecies justice.

Accordingly, I argue that education, healthcare, and food systems should each be structured by distinct constitutional principles grounded in care and solidarity, and that these principles ought to guide public policy within their respective spheres.

4 Education as Constitutional Practice

Constitutions frequently address education because it is one of the primary institutional sites through which a polity reproduces itself. The right to education is constitutionally protected in numerous jurisdictions, including Germany, India, and Chile, where education is framed either as a fundamental right or as a core state responsibility. Such provisions reflect a structural commitment to cultivating the capacities upon which democratic legitimacy depends.

If self-development is treated as a foundational constitutional value, then the constitutional status of education cannot be confined to human advancement alone. This does not entail that the state must "educate" nonhuman animals; rather, it requires that public education cultivate just forms of human–animal relations so that animals are not impeded in exercising their own capacities for self-development. Within a multispecies constitutional order, education functions as a central mechanism for dismantling speciesist norms and cultivating dispositions of care and solidarity capable of resisting institutional forms of interspecies oppression.

Seen from this perspective, constitutional attention to education is not incidental. It is an institutional expression of care: a commitment both to supporting the self-development of members of the political community and to sustaining the relational conditions under which self-development – human and nonhuman – becomes possible.

4.1 From Instruction to Ethical Formation

This ambition requires conceiving education not merely as formal instruction but as a relational and ethical practice through which citizens are formed and political norms are reproduced. What this underscores is the educational system's immense capacity to shape interpretive frameworks, moral boundaries, and the scope of perceived obligation. A caring constitution would therefore foreground education as a formative site in which empathy and moral imagination are cultivated as politically indispensable capacities.[3]

Read constitutionally, this is a requirement of legitimacy. A polity that systematically educates its members to ignore or normalise institutionalised suffering cannot plausibly claim to be organised around justice.

Importantly, this claim does not require that animals be recognised as citizens in the sense defended by Donaldson & Kymlicka (2011) in *Zoopolis*. The claim is more modest and, in some respects, more demanding: a just state must recognise animals as morally relevant cohabitants of the political community. In practice, this means that their capacities for self-development – and the institutional conditions that frustrate or sustain those capacities – must inform constitutional commitments in education, law, and public policy.

4.2 The Curricular Framing of Nature

If education is constitutive of political community, school curricula become constitutionally relevant sites of normative formation. Contemporary curricula frequently reproduce human-centred logics, often reflecting technocratic models of knowledge that naturalise extractive relations with the more-than-human world.

A revealing example appears in Chile's 4th-grade national social sciences curriculum, which includes the objective of recognising and locating the main natural resources of the Americas, considering their geographical distribution, use, and the importance of caring for them within the framework of sustainable development (Ministerio de Educación de Chile, 2023). Accompanying materials list oil, zinc, and lead alongside coffee, cattle, and salmon, all framed abstractly as "resources" for human use.

This framing collapses morally distinct entities into a single economic category. The extraction of lead appears conceptually continuous with the breeding and slaughter of calves. Sentient beings are rendered equivalent to mineral deposits. Such curricular structures do not merely transmit information; they encode a political ontology in which nonhuman life is positioned as instrumentally available. A caring constitution must therefore recognise that the right to education cannot be neutral with respect to the ethical framing of nature.

As Rachel Carson (1962) and Aldo Leopold (1949) argued decades ago, the ecological crisis is inseparable from a crisis of cultural imagination. Education plays a central role in reproducing the metaphysical assumptions that naturalise the instrumentalisation of nonhuman life.

4.3 School Food and Interspecies Justice

The educational sphere also includes institutional practices often treated as peripheral to pedagogy – most notably school meal provision. Yet in a multispecies polity, food policy within schools is constitutionally significant.

National school meal programmes are typically justified in terms of child welfare and equality. However, they frequently rely on and normalise industrial animal agriculture. In doing so, they embed ecological degradation and interspecies violence within the everyday experience of state-supported education.

Livestock today accounts for roughly 60 per cent of global mammalian biomass while wild mammals constitute only approximately 4 per cent (Bar-On et al., 2018). Domestic poultry similarly outweighs wild bird populations by more than two to one. This dramatic reconfiguration of planetary biomass reflects not an ecological anomaly but a political economy in which sentient life is systematically organised around commodification.

Food cannot be regarded as nutritionally neutral. It is politically, ecologically, and morally charged. Industrial animal agriculture contributes to climate breakdown, biodiversity loss, and zoonotic risk, while simultaneously institutionalising routinised violence against sentient beings (Schlottmann & Sebo, 2019; Rowlands, 2021). A school system organised around interspecies justice cannot treat such practices as ethically invisible.

This does not imply that constitutional law should dictate menus in detail. Rather, it requires that constitutional principles of care and solidarity inform how educational institutions conceptualise nourishment, sustainability, and relational responsibility. Educational environments should not reproduce structural injustice under the guise of protein delivery or efficiency metrics.

4.4 Education and the Moral Boundaries of the Polity

Education, then, is not merely a vehicle for knowledge transmission. It helps constitute the moral boundaries of the political community. It shapes how harm is interpreted, which beings are recognised as vulnerable, and whose capacity for self-development is regarded as a matter of collective responsibility.

A human-centred legal imaginary can be reproduced through education, but it can also be transformed. A caring constitution requires that educational commitments cultivate an interspecies civic ethos – one capable of apprehending animals as subjects of vulnerability and interest rather than as mere resources or background conditions for human progress.

By recognising education as a domain of ethical and political formation, multispecies constitutionalism expands the scope of justice beyond the human. It reconfigures state responsibility in light of interdependence and shared vulnerability, embedding interspecies responsibility not only through formal legal provisions but through the pedagogical practices that sustain and legitimise those provisions over time.

The foregoing analysis may be crystallised in the following constitutional principle of interspecies ethical formation: *the state shall organise public education in a manner consistent with cultivating care, solidarity, and non-oppressive relations toward all self-developing beings affected by its institutions.*

5 Healthcare

Many constitutions impose positive obligations on the state to organise public healthcare systems, ensure equitable access, prevent disease, and protect population health (e.g., Matsuura, 2015; Giovanella et al., 2019). These commitments reflect a foundational constitutional insight: health is not merely a private good but a precondition of full participation in political life. A polity that fails to secure the basic health conditions of those subject to its authority undermines its own legitimacy, because exposure to illness, injury, and dependency is not an accidental misfortune but a structural feature of shared life.

Viewed through a multispecies lens, this logic cannot remain anthropocentric. Human political, economic, and legal institutions systematically shape the health conditions of nonhuman animals. Industrial agriculture, confinement systems, habitat destruction, selective breeding, and pollution produce predictable patterns of disease, injury, and premature death across species. These harms are not incidental; they are institutionally generated. Where animals are self-developing beings whose bodily integrity and developmental capacities are routinely structured by institutional design, constitutional responsibility extends to the health conditions those institutions create.

This responsibility entails the constitutional recognition of animal health through both negative constraints on practices that predictably undermine animals' health and positive duties to organise human–animal relations in ways compatible with care and solidarity. Health, on this account, functions as a normative threshold conditioning the legitimacy of state action rather than as an instrumental good subordinate to human interests. This moves beyond animal welfarism by framing animal health in terms of enforceable entitlements: entitlements to protection against preventable harms and, where appropriate, to forms of care and assistance that sustain basic functioning within shared socio-ecological systems.

Contemporary policy discourse has sought to approximate aspects of this vision through the increasingly influential One Health framework, which emphasises the interdependence of human, animal, and environmental health. While this approach usefully disrupts siloed governance models, it remains normatively limited. In practice, One Health initiatives frequently justify animal health interventions in instrumental terms – disease spillover prevention, economic resilience, or food security – rather than as obligations grounded in justice toward animals themselves.

Consider large-scale responses to avian influenza outbreaks (Donoso & Gimpel, 2024). Millions of poultry are routinely culled pre-emptively in the name of public health protection and economic stability. These measures may be epidemiologically justified within existing systems, yet they treat animals solely as vectors of risk or units of production. No constitutional question is raised about the legitimacy of institutional arrangements that systematically generate the conditions under which such mass killing becomes routine. One Health, in this configuration, mitigates harm without interrogating its structural production.

One Health therefore lacks the conceptual resources to articulate stable entitlements, impose robust constitutional constraints on institutional design, or secure the standing of animals within public health governance as justice-bearing subjects.

This limitation becomes visible even in comparatively progressive initiatives. In South Sudan, *Vétérinaires Sans Frontières* has developed a Community-based Animal Health Worker model to deliver veterinary services in remote, conflict-affected regions (VSF International, n.d.). By training members of livestock-keeping communities in basic animal healthcare and disease prevention, the programme demonstrates the practical feasibility of integrating animal health into broader public health strategies under conditions of extreme vulnerability. Yet its reliance on emergency logics, productivity concerns, and contingent funding underscores the fragility of approaches that lack constitutional grounding. What appears as care remains structurally conditional rather than institutionally secured.

Such initiatives should not substitute for justice but clarify what justice requires. Integrating animals into public health demands more than intersectoral coordination; it requires recognising animal health as a constitutional concern grounded in principles of care, solidarity, and institutional responsibility.

Preventive health policy must therefore address the structural conditions – industrial agriculture, habitat destruction, and pollution – that systematically generate disease and suffering across species. Health, on this account, is relational: not merely the absence of pathology but the presence of social and ecological systems that sustain the capacities for self-development of affected beings. A polity guided by care measures its success not only by human health outcomes but by the vitality of the multispecies communities within which political life unfolds.

If health conditions the legitimacy of political authority, exposure to preventable disease, injury, and institutionalised bodily harm cannot be systematically imposed on self-developing beings subject to state power. A multispecies constitutional order should therefore recognise the following principle:

The state shall organise economic, environmental, agricultural, and public health systems to prevent structural exposure to disease and to preserve the bodily integrity and developmental capacities of all self-developing beings subject to its authority.

This principle does not mandate identical health regimes across species. It establishes instead a threshold of legitimacy: institutional arrangements that predictably generate widespread suffering, pathological confinement, or preventable disease cannot be justified as neutral exercises of public authority. Health, understood relationally, thus operates as a constitutional constraint on the organisation of shared life.

6 Systems of Food Production

Unlike education and healthcare, food systems do not typically appear as separate chapters in constitutions. It has been observed that food is rarely treated as a central constitutional theme, despite its potential to improve food governance and access (Adams, 2021; Sampson et al., 2021). Where food is explicitly mentioned, it is usually framed through the right to food, particularly in provisions concerning children, nutrition, or minimum standards of living, as illustrated by examples from Brazil, Colombia, Haiti, Malawi, and Ukraine. In such contexts, food appears primarily as an entitlement to adequate nutrition rather than as a structured system of production, distribution, and power.

Yet even when food systems are not textually constitutionalised, they are deeply embedded in constitutional orders. Outside the constitutional text, food is extensively regulated through ordinary law: agricultural policy, land tenure regimes, trade agreements, subsidies, environmental regulation, animal welfare statutes, public procurement rules, and public health legislation all shape how food is produced and distributed. These legal frameworks are themselves authorised and constrained by constitutional principles governing property, economic freedom, environmental protection, and state responsibility. In this sense, food systems are already constitutionally mediated, even if not named in constitutional documents.

The constitutional relevance of food systems therefore does not depend on explicit textual inclusion but on their structural role in sustaining political sovereignty and collective life. A political community's capacity for self-determination presupposes secure and just access to nourishment. Food sovereignty, often invoked in constitutional and political discourse, is inseparable from the organisation of food systems that determine how food is produced, distributed, and accessed. To treat these systems as constitutionally peripheral is to overlook their foundational role in political authority itself.

From the standpoint of interspecies justice, food systems occupy a central institutional position. They are among the most consequential sites through which hierarchical and oppressive relations between humans and nonhuman animals are structured and sustained. Industrial animal agriculture systematically subordinates animal lives to economic efficiency, profit maximisation, and consumption demands. It produces structural vulnerability, routinised violence, and large-scale confinement regimes that predictably undermine animals' bodily integrity and capacities for self-development.

A constitution grounded in care and solidarity cannot treat such systems as morally neutral or as matters of private preference. Where institutional arrangements entrench avoidable harm and oppression, they fall within the scope of constitutional concern. Food production and distribution are therefore not merely economic sectors; they are mechanisms through which political authority organises relations of vulnerability and power across species.

A caring constitution must therefore recognise the following principle of multispecies food justice:

The state shall organise food systems in ways consistent with the interests and self-development of all beings affected by them.

This principle does not constitutionalise specific dietary prescriptions. Rather, it establishes food production and distribution as domains of public responsibility in which state duties extend beyond human nutrition to encompass the conditions under which animals live and develop. It supplies interpretive guidance for legislation, agricultural regulation, and adjudication, requiring that efficiency, profitability, or cultural preference not automatically override considerations of animal integrity and justice.

At a minimum, such a principle entails three normative commitments:

1. Negative duties: The state must prevent and dismantle institutionalised practices that predictably undermine animals' bodily integrity and capacities for self-development, including intensive confinement, genetic manipulation oriented solely toward productivity, and large-scale industrial slaughter where these practices constitute structural injustice.
2. Positive duties: The state must foster food systems that do not depend on systemic exploitation. This requires sustained support for plant-based

production, agroecological practices, biodiversity protection, and deliberate transitions away from exploitative agricultural models. A substantial reallocation of public subsidies from industrial animal agriculture toward just and sustainable food systems is imperative. Public procurement – across schools, hospitals, prisons, and other state institutions – emerges as a central mechanism for institutionalising these commitments in practice.
3. Interpretive orientation: All food-related legislation and policy must be interpreted in light of interspecies care and solidarity rather than treated as technically neutral domains governed exclusively by market efficiency or productivity.

Under this framework, food sovereignty itself must be reinterpreted. While typically framed in terms of local control and cultural continuity, sovereignty cannot legitimately extend to practices that impose systematic and avoidable oppression on nonhuman beings. Collective self-determination is bounded by duties of care and reciprocity toward those whose lives are structured by that collective.

This claim requires a qualification. Chapter 5 showed how the imposition of dominant economic norms on Indigenous communities constitutes a form of cultural imperialism – the universalisation of one group's perspective at the expense of others' ways of life. A similar objection might be raised against constitutional constraints on food practices: that limiting culturally embedded food traditions in the name of interspecies justice reproduces the very logic of cultural oppression the book seeks to contest, particularly where those traditions belong to communities already subject to colonial dispossession.

The tension is real and cannot be dismissed. Three points, however, suggest it does not defeat the argument. First, the constitutional principle proposed here does not target culturally specific food practices as such; it targets the institutional architecture of industrial animal agriculture – a system that is itself a product of colonial and capitalist expansion rather than of traditional or Indigenous food cultures, and one that has in many cases displaced rather than expressed those cultures. Second, the principle imposes constraints on state authority and public institutions rather than on individual or communal practice – it governs what the state may fund, subsidise, and procure, not what citizens or communities may eat. Third, the framework of care and solidarity that grounds the principle is not presented as a culturally neutral universal but as a critical response to the specific forms of structural oppression analysed in Part I (see especially Chapter 5 on Indigenous communities and global commodity chains). The application of this framework must therefore be attentive to the structural position of the communities whose practices it affects.

Crucially, constitutional attention to food systems presupposes recognising animals as subjects of justice within those systems. Once animals are no longer treated merely as commodities, their interests acquire direct normative relevance in agricultural regulation (see Chapter 8). The justificatory structure of food policy is thereby transformed: what was previously defended as market demand or cultural preference becomes subject to constitutional scrutiny.

Food policy is not a technically neutral arena. Where public institutions permit food systems that systematically generate suffering, dependency, and premature death, they entrench forms of oppression incompatible with interspecies justice. Conversely, a constitution that assumes responsibility for the ethical architecture of food production affirms that political authority is accountable not only to human inhabitants of the political order but also to the nonhuman animals whose lives it organises, constrains, and often extinguishes.

The agricultural food system therefore emerges not as a peripheral policy domain but as a central pillar of a multispecies polity. A caring constitution reconfigures food governance as a core site where commitments to care and solidarity are realised.

7 Conclusion

The constitutionalisation of interspecies care and solidarity does not, by itself, guarantee justice for animals – just as constitutional protections for human rights or environmental integrity do not automatically secure dignity or ecological balance. Constitutions can articulate ethical aspirations and structure public authority, but they cannot alone generate the dispositions required to sustain them. Animals may be recognised as subjects of justice, yet such recognition risks remaining inert unless embodied in everyday practices that reshape how citizens perceive, feel, and act within the shared world.

A polity grounded in care and solidarity demands more than institutional reform. It requires the gradual transformation of the civic dispositions through which citizens perceive authority, relate to others, and interpret what it means to belong. In most contemporary societies, those dispositions remain deeply anthropocentric – presupposing a sharp moral divide between humans and other animals, normalising indifference to nonhuman suffering, and rendering structural exploitation unremarkable. What is required instead is an interspecies civic ethos: a way of inhabiting shared life that recognises other animals as co-participants in a common world rather than as instruments, threats, or background scenery.

Such a transformation cannot be decreed by constitutional text. It must be cultivated through the ordinary sites in which moral imagination is formed and reproduced. Schools and universities must foster attentiveness, empathy, and critical reflection on human–animal relations. Urban planning can enable new forms of coexistence through wildlife corridors, animal-friendly public spaces, and ecological integration. Public procurement, agricultural policy, and health governance can reconfigure the material conditions under which animals live and die. These interventions reshape the visible and invisible structures of everyday life.

When citizens learn to see pigeons not as pests but as cohabitants; when transport systems are designed to prevent roadkill; when students debate animal ethics and food systems as matters of justice rather than lifestyle preference – a new political subjectivity begins to emerge. Care becomes not merely a private virtue but a public competence, part of what it means to participate responsibly in a shared political order.

In this light, the constitution functions not only as a legal instrument but as a normative architecture. By naming interspecies justice as a constitutional concern, it authorises the transformation of political imagination itself. It does not presume an already caring citizenry; it establishes the institutional conditions under which care and solidarity can become durable features of civic life.

Yet this project remains imperfect and incremental. Progress toward multispecies justice will unfold unevenly, through partial reforms and contested reinterpretations. Modest institutional shifts – redesigning public spaces to accommodate wildlife, embedding interspecies ethics within educational curricula, phasing out animal exploitation from public procurement – may appear limited in isolation. But cumulatively, they can reorient the moral grammar of political life, cultivating the civic orientation that Iris Marion Young describes as attentiveness to structural injustice and relational responsibility.

The following chapters examine this transformation at a more concrete level. While this chapter has articulated the constitutional horizon of a caring multispecies polity, the next two explore how that horizon reshapes the spaces in which political life is lived daily: the food system and the city. It is there – within urban infrastructures and agricultural economies – that the principles of care and solidarity are either embodied or betrayed.

Notes

1 Article 71 of the Ecuadorian Constitution recognises the right of any individual or collective to demand the protection and reparation of Nature. In Sentencia No. 253-20-JH/22, the Constitutional Court interpreted this provision to include constituent animals within Nature's protected integrity (Constitutional Court of Ecuador 2022, p. 161).

2 Article 20a, Basic Law for the Federal Republic of Germany (1949, amended 2002): "Mindful also of its responsibility toward future generations, the state shall protect the natural foundations of life and animals by legislation and, in accordance with law and justice, by executive and judicial action, within the framework of the constitutional order."

3 When thinking about empathy, I have in mind Lori Gruen's idea of *entangled empathy*, a form of perception focused on attending to another's experience – a process involving emotion and cognition in which we recognise we are in relationships with others (Gruen, 2015, p. 3).

References

Adams, F. (2021). *The right to food. The global campaign to end hunger and malnutrition*. Springer.

Animal Welfare Board of India v. A. Nagaraja. (2014). Civil Appeal No. 5387 of 2014. Supreme Court of India.

Bar-On, Y. M. et al. (2018). The biomass distribution on Earth. *Proceedings of the National Academy of Sciences*, 115(25), 6506–6511.

Carson, R. (1962). *Silent spring*. Houghton Mifflin.

Constitutional Court of Ecuador. (2022). *Judgment No. 253–20-JH/22: Case of Estrellita*. Corte Constitucional del Ecuador.

Di Concetto, A. (2017). « Desordem e Regresso »: Évolution de la jurisprudence constitutionnelle de la Cour Suprême brésilienne en matière de protection de l'animal. *Revue semestrielle de droit animalier*, 2, 465–492.

Donoso, A. (2021). Toward a new framework for rights of the biotic community. In D. Corrigan & M. Oksanen (Eds.), *Rights of nature: A re-examination* (pp. 140–155). Routledge.

Donoso, A., & Gimpel, J. (2024). Dimensiones éticas del sacrificio sanitario de animales en crisis epidémicas: El caso de la influenza aviar en Chile. *Temas de la Agenda Pública*, 19 (174), 1–14. https://repositorio.uc.cl/server/api/core/bitstreams/b5780831-71b9-487b-a8b2-87f3b740d130/content.

Donaldson, S., & Kymlicka, W. (2011). *Zoopolis: A political theory of animal rights*. Oxford University Press.

Sampson, D. et al. (2021). Food Sovereignty and Rights-Based Approaches Strengthen Food Security and Nutrition Across the Globe: A Systematic Review. *Frontiers in Sustainable Food Systems*, 5. doi:10.3389/fsufs.2021.686492.

Matsuura, H. (2015). State constitutional commitment to health and health care and population health outcomes: evidence from historical US data. *American Journal of Public Health*, 105 (Suppl 3), e48–54. doi:10.2105/ajph.2014.302405.

Giovanella, L. et al. (2019). De Alma Ata a Astaná. Atención primaria de salud y sistemas universales de salud: compromiso indisociable y derecho humano fundamental. *Cadernos de saude publica* 35 (3), e00012219. doi:10.1590/0102-311x00012219.

Gruen, L. (2015) *Entangled Empathy. An Alternative Ethic for Our Relationships With Animals*. Lantern Books.

Kansal, V. (2016). The curious case of Nagaraja in India: Are animals still regarded as "property" with no claim rights? *Journal of International Wildlife Law & Policy*, 19 (3), 256–265.

Leopold, A. (1949). *A Sand County almanac*. Oxford University Press.

Ministerio de Educación de Chile. (2023). *Bases curriculares: Ciencias sociales, 4° básico*. Ministerio de Educación, Gobierno de Chile. www.curriculumnacional.cl.

Montes Franceschini, M., & Stilt, K. (2023, February 10). Estrellita the woolly monkey and the Ecuadorian Constitutional Court: Animal rights through the rights of nature. *ReVista: Harvard Review of Latin America*. https://revista.drclas.harvard.edu/estrellita-the-wooly-monkey-and-the-ecuadorian-constitutional-court-animal-rights-through-the-rights-of-nature/.

Rowlands, M. (2021). *World on Fire: Humans, Animals, and the Future of the Planet*. Oxford University Press.

Schlottmann, C., & Sebo, J. (2019). *Food, animals, and the environment: An ethical approach*. Routledge.

VSF International. (n.d.). *Community-based animal health workers*. Vétérinaires Sans Frontières International. https://vsf-international.org/themes/cahws/.

8

THE ARCHITECTURE OF A JUST FOOD SYSTEM

1 Introduction

In a 2015 opinion article, Yuval Noah Harari argued that animals are the main victims of our time and that industrial farming constitutes perhaps the worst crime ever committed, citing systematic confinement, bodily mutilation, forced separation of offspring, and genetic manipulation that inflicts profound suffering (Harari, 2015). A few years earlier, a similar reflection led Kwame Anthony Appiah to imagine future generations looking at factory farming with the same moral incomprehension now reserved for slavery and colonialism, asking what people could possibly have been thinking (Appiah, 2010). Both thinkers express the profound ethical horror at the core of contemporary food systems, in which routinised and largely invisible forms of violence have been normalised in industrialised societies.

The only adequate response to this situation is a radical reorientation of industrial food systems. These systems – spanning production, processing, distribution, consumption, and waste – are increasingly coordinated by transnational agribusiness and capital-intensive methods that optimise efficiency, standardisation, and profit at the expense of ecological and physiological flourishing (Clapp, 2020; Clark et al., 2021). This configuration produces intertwined social, political, ethical, and ecological crises: labour precarity and exposure to toxic inputs; concentrated corporate power and marginalisation of small producers (Holt-Giménez, 2017; Howard, 2016); pervasive and routinised animal suffering (Singer, 2009; Nussbaum, 2006; Twine, 2021); and accelerated ecological degradation, including climate disruption, deforestation, biodiversity loss, soil depletion, and water scarcity (Steinfeld et al., 2006; IPCC, 2019; Weis, 2013).

This chapter examines how food systems might be reconfigured through principles of care and solidarity, understood as normative commitments that

DOI: 10.4324/9781003434856-10

shape institutions. Against dominant models predicated on extraction, disposability, and detachment, a care-based food framework treats feeding and being fed as morally and politically charged relations that bind human and nonhuman lives within networks of interdependence.

To develop this vision, the chapter highlights emerging practices that prefigure alternative futures: agroecological systems fostering biodiversity and ecological resilience; plant-based innovation decoupling nutrition from animal use; cultivated meat technologies aiming to eliminate slaughter; and non-extractive multispecies farming that recognises animals as co-inhabitants rather than commodities. These initiatives demonstrate that care can be materially enacted and institutionalised, and that a just food system is not merely aspirational but already operative, if marginalised, as a form of food production.

What follows is organised in four parts. First, I present a normative framework of interspecies care and solidarity that distinguishes the chapter's position from welfarism and from existing abolitionist and rights-based alternatives. Second, I analyse agri-food practices that embody this orientation through relational ethics and coexistence. Third, I develop the institutional and political conditions necessary for a post-oppressive food system, showing why reforms internal to the industrial paradigm are frequently absorbed as symbolic adjustments rather than genuine transformation. Fourth, I identify the governance mechanisms through which care and solidarity can be structurally embedded across the food system as a whole.

2 Care, Not Animal Welfare

If a food system is organised around extraction, disposability, and the effacement of moral relations, its alternative cannot simply be a milder iteration of the same oppressive logic. A just food system does not inflict less harm within an unchanged structure. Instead, it refuses to institutionalise harm as a condition of production. The task is not to humanise existing arrangements but to reconfigure them around care and solidarity as foundational principles that acknowledge animals as subjects of justice rather than optimisable units of output.

Care should not be understood as an ethical supplement to individual food choices but as a structural requirement to secure the conditions under which vulnerable beings can develop. To employ care in the governance of food systems is not to soften supply chains but to redefine permissible institutional relations. Practices are not justified by merely reducing suffering but by upholding the moral standing of those involved. This means rejecting the current food system's arrangements predicated on instrumentalisation and disposability.

This shift from welfare to care should not be confused with existing abolitionist or rights-based positions, though it shares important ground with both.

Peter Singer's utilitarianism, the most influential consequentialist framework in animal ethics, correctly identifies animal suffering as morally significant and industrial agriculture as a moral catastrophe. Yet its calculative structure treats attentiveness, responsiveness, and relational orientation as instrumentally rather that constitutively valuable, missing the relational substance of care. Furthermore, Singer's analysis operates primarily at the level of individual moral theory rather than at the structural and institutional failures that this book addresses.

Rights-based frameworks, most prominently Tom Regan's, supply stronger protections by grounding animal claims in inherent worth rather than aggregate welfare. Care does not compete with these frameworks but complements them: where rights specify what must not be done to animals, an interspecies politics of care specifies what institutions must actively provide – the relational conditions, attentive practices, and material assurances under which self-development becomes possible. Gary Francione's abolitionism comes closest to the position defended here in its rejection of welfare reform as moral legitimation of oppression. But Francione's view that ethically just relations require the abolition of domestication – his insistence that all use is exploitation and that the only just relation is independence – forecloses precisely the forms of non-oppressive cohabitation, mutual dependency, and care-inflected participation that this book's framework requires. The position here is not that human-animal relationships must end but that they must be transformed: from relations of instrumentalisation and disposability into relations of care, solidarity, and shared self-development.

Welfare-based approaches to animal agriculture are therefore inadequate for the task. They leave unquestioned the legitimacy of systems that rely on animal exploitation, ecological extraction, and labour precarity, treating harms as adjustable variables rather than as consequences of an unjust structure. When institutional design systematically negates agency, degrades ecosystems, or normalises exploitation, incremental mitigation operates within a context of injustice rather than against it. Making structural harm more tolerable refines injustice rather than overcoming it, and allows welfare to function as moral legitimation of oppression rather than as a genuine challenge to it.

This shift toward a politics of care and solidarity also unsettles the capitalist displacement of responsibility onto individual consumers. Structural harms arise from the routine operation of social institutions, not from isolated choices. Contemporary capitalist food regimes moralise individual consumption and privatise responsibility (Shamir, 2008; Brown, 2015), obscuring the real source of widespread harm. The harms of industrial food systems stem not primarily from personal dietary decisions but from institutional architectures – subsidy regimes, property relations, trade rules, liability frameworks, and procurement standards – that systematically privilege efficiency and profit.[1]

The emphasis must therefore fall on collective institutional agency rather than on the ethical consumer's cultivated virtue. Solidarity is not a discretionary moral sentiment but a political obligation to reconfigure the structures that organise and distribute wrongful harm (Twine, 2010). In the context of food systems, this means directing political energy not toward individual dietary choices but toward the institutional arrangements – subsidies, procurement rules, property regimes, and trade agreements – that make exploitation structurally default and care structurally exceptional.

Recognising animals as subjects of justice consolidates this perspective. Attentive recognition of their vulnerability, agency, and capacity for self-development establishes moral constraints on permissible food practices. Reducing animals to commodified means of profit extraction places them outside the bounds of justice regardless of welfare improvements or technological sophistication. Because institutional moral legitimacy hinges not on market performance but on the justifiability of practices to those affected by them – including nonhuman beings historically excluded from normative consideration – current industrial animal agriculture systems must be regarded as lacking moral legitimacy.

The proposed approach also centres on ecological integrity. Food systems are ecological assemblages that entangle human and nonhuman lives in shared conditions of survival. A care-based framework demands ecological co-viability rather than sustainability appended to extractive regimes. Externalising ecological harm undermines the very conditions upon which food systems depend and thereby violates the core commitments of care. A genuinely non-anthropocentric sustainability must be constitutive of institutional design, not appended to practices whose essence remains extractive.

Care must therefore be embedded in the material and institutional architecture of food systems. It is insufficient to exhort individuals to withdraw from unjust practices when those practices remain structurally default. A just food regime must be configured such that disposability and ecological harm are not routine options. This requires comprehensive institutional transformation.

These considerations clarify why incrementalist strategies that seek to humanise the existing regime are inadequate.[2] Such strategies attempt to ameliorate symptoms without altering the structural conditions that produce harm (Sciortino et al., 2025). Programmes such as Certified Humane, Whole Foods' 5-Step Animal Welfare Rating, high-welfare labels, and mainstream organic certification serve not as paths to justice but as mechanisms of legitimacy management. As Twine (2010) and Giraud (2021) have shown, these labels enable consumer moral outsourcing, in which the act of purchasing becomes a simulation of ethical engagement rather than a lever of transformation.

Similarly, so-called regenerative agriculture, increasingly adopted by major food corporations, often operates within the very logics of extractivism it claims to reject. Despite promising ecological restoration, these initiatives frequently lack rigorous criteria and allow continued confinement, mutilation, and slaughter under the banner of sustainability (de la Bellacasa, 2017; Chiles,

2023). When regenerative programmes are decoupled from animal justice, they function as greenwashed speciesism.

Such reforms do not dismantle oppression; they rebrand it. They displace the most egregious inputs – battery cages, routine antibiotic use – but preserve the underlying relations of instrumentalisation and capital concentration. The industrial food system is adaptive, absorbing critique not as a threat but as a reputational asset.

A care- and solidarity-based approach treats justice as a constitutive limit of the food system. Practices that presuppose the disposability or pure instrumentality of animals are excluded from legitimacy irrespective of efficiency or profitability. A system premised on confinement, ecological depletion, and labour precarity cannot be rendered just through marginal welfare gains; these are not incidental properties but structural features of the regime. The relevant task is not to calibrate acceptable levels of harm but to design institutions in which systemic harm is no longer a condition of functioning. A system that requires oppression to operate must be replaced, not optimised.

3 Just Food Systems

3.1 Counter-Regimes and Their Normative Significance

Across contemporary food economies, there exist practices that depart from the industrial regime not merely in method but in their underlying relational and ethical logic. Community-supported agriculture, cooperative food networks, and solidarity purchasing schemes rely on reciprocal relations rather than anonymous market exchange, distributing responsibility between producers and consumers and thereby materialising a principle of shared vulnerability. Agroecology, regenerative agriculture, and multispecies farming enact ecological co-viability rather than yield-maximisation, treating animals – when present – not as interchangeable inputs but as agents whose lives and movements shape decisions on the land. Plant-based systems and institutional plant-forward procurement instantiate care by eliminating rather than diluting the exploitation of animals, while novel protein technologies such as precision fermentation and cellular agriculture achieve the same displacement by technological means.

What is normatively significant about these arrangements is not their current scale but their structural orientation: they instantiate forms of feeding that cannot be reconciled with the commodity logic of confinement and exploitation. They serve as empirical counter-evidence to the claim that the industrial regime is necessary: they materialise care precisely where dominant capitalist logic declares care infeasible. Their marginality is not a defect but the outcome of the current regime's design, which renders ethical provisioning structurally exceptional. The existence of such counter-regimes does not weaken the indictment of the industrial system but sharpens it – showing that injustice persists not for lack of workable alternatives but because the prevailing order actively prevents their expansion.[3]

3.2 Agroecological and Integrated Systems

One of the most promising institutional models is agroecology. Defined as the application of ecological principles – species interaction, biodiversity dynamics, and ecosystem resilience – to agricultural design and management, agroecology promotes inclusive and long-term sustainability (Altieri, 1995; Altieri & Nicholls, 2005; FAO, 2019).

Ivette Perfecto and John Vandermeer illustrate this through coffee plantations, showing how intercropping coffee with diverse shade trees creates habitats for a wide array of species, enhancing biodiversity and naturally regulating pests (Perfecto & Vandermeer, 2015; Philpott & Bichier, 2012). Their analysis, they argue, holds broader significance: the ecological functioning of the coffee agrosystem generates implications for all human-dominated ecosystems in the tropics (Perfecto & Vandermeer, 2015, p. 8). Agroecology thus models a form of interspecies collaboration that moves toward a general relational, non-oppressive agriculture.

This vision is further developed in Perfecto et al. (2019), where similar frameworks are applied to cacao, maize, and rice systems. The authors foreground the political and economic barriers that inhibit the expansion of low-intensity, biodiversity-enhancing farming, arguing that small-scale traditional practices are being systematically undermined by global agricultural regimes that prioritise industrial-scale, high-input monocultures.[4] Viable agroecological approaches therefore remain politically marginalised – revealing the structural limits of conventional agricultural policy, which continues to prioritise productivity and output growth over ecological sustainability and social justice. What is required is a paradigmatic shift toward integrated frameworks that reconcile economic viability with ecological resilience and democratic governance (Siebrecht, 2020).

The agroecological production of rice through integrated rice-duck and allied paddy polycultures offers a particularly illuminating example. When introduced into flooded paddies, ducks reduce weeds and crop-damaging insects, oxygenate the water through their movement, and contribute manure that enhances soil fertility – improving or maintaining yields while reducing or eliminating synthetic inputs (Wu & Liu, 2024; Zhang et al., 2023; Ghaznavi et al., 2025).

The ecological efficiency of these systems matters less than what they reveal about the contingency of the extractive endpoint. In conventional versions, this symbiosis is folded back into an extractivist logic: ducks function as living tools for pest control and are then slaughtered or their reproductive capacities appropriated as an additional profit stream. Yet the ducks' contribution to the agroecosystem – regulating pests, oxygenating water, enriching soil – is complete without their subsequent slaughter. That endpoint is not a feature of the agroecological relationship but an addition imposed by the logic of capital maximisation.

These cases are therefore best understood as proof of concept: they demonstrate that non-oppressive human-animal cooperation in food production is not utopian but already operative, requiring not invention but political will to strip away the extractive layer that capital has deposited on top of it. Re-imagined from the standpoint of animals as subjects of justice, ducks can be recognised as co-labourers who inhabit and reshape the agroecosystem rather than as disposable biotechnologies. What emerges is not a hierarchy of beings but a multispecies assemblage in which mutual exchanges sustain the flourishing of the whole.

This reframing echoes Iris Marion Young's (2000, especially chapter 2) critique of communicative norms presented as universally valid yet structured by the interests and capacities of dominant groups. Just as Young shows how apparently neutral procedural standards encode the perspectives of those who designed them, the extractive endpoint of conventional agriculture presents itself as an economic necessity while reflecting the priorities of capital rather than the requirements of the agroecological relationship itself. Agroecological practices, by making this contingency visible, function as experimental political arrangements – limited yet meaningful efforts to materialise multispecies justice within concrete institutional settings, and to demonstrate that the dominant logic is a choice rather than a constraint.

A parallel, though differently configured, departure from extractive agriculture can be observed in integrated farming systems in the wetland ecosystems of Bangladesh. Where the rice-duck case shows that non-extractive cooperation is *fully* realisable, the Bangladeshi systems show how integrated arrangements can move *toward* non-extractive relations while still operating within commercial agriculture. Here, farmers integrate multiple enterprises – crops, livestock, fish farming, and agroforestry – not for maximised extraction but for closed-loop sustainability, diversification, and livelihood security. A recent study has shown that the most successful systems combine homestead agroforestry, livestock, and open-water fish catching not as isolated economic activities but as overlapping ecological functions, where outputs from one system support the others (Akter et al., 2025). While animals in these systems are still partially commodified, their role is reframed through patterns of mutual productivity, nutrient cycling, and long-term care commitments – and, as with the rice-duck case, the partial commodification is an addition to rather than a requirement of the underlying ecological logic.

Akter et al. (2025) note not only increases in income but transformations in farmers' relationships to land, animals, and ecological knowledge, particularly through training, community solidarity, and the redistribution of material and symbolic capital. These multispecies assemblages suggest a route by which agriculture could move closer to multispecies justice: not necessarily by removing animals from agricultural relationships entirely, but by transforming those relationships into care-inflected participation integrated across ecological, economic, and social dimensions. The extractive endpoint is neither inherent nor necessary.

3.3 Plant-Based and Cellular Systems: A Plural Architecture

At the other end of the spectrum, emerging plant-based and cellular food technologies instantiate care not through integration but through removal. Plant-based and cultured meat systems demonstrate that the nutritional and gustatory goods for which animals are exploited can be achieved without subjugating them to structures that deny agency and flourishing. Cultured meat – animal flesh produced by growing animal cells in bioreactors rather than raising and killing animals – illustrates this transformation most clearly. The normative significance of the technology lies not in its novelty but in its potential to render oppression unnecessary. If the same nutritional outcomes can be achieved without confinement, forced reproduction, or slaughter, then the continuation of those practices is not inevitable but elective.

These innovations are not speculative. El Wali et al. (2024) show that using renewable energy sources, up to 72 per cent of global livestock products could be replaced by cell-based proteins by 2050 – a model suggesting that cultured meat could supplement or largely replace conventional animal production while drastically reducing land use, greenhouse gas emissions, and water consumption. When combined with dietary shifts toward plant-based nutrition, the results could sustain the caloric needs of ten billion people without relying on the systematic exploitation of animals or the destruction of ecosystems (Willett et al., 2019).[5] These projections align with the EAT-Lancet Commission's call for a comprehensive reorganisation of food systems to meet both planetary and public health thresholds (Willett et al., 2019, p. 448).

These possibilities – agroecology, plant-based systems, cellular agriculture, fermentation-based proteins – are not competing solutions but complementary elements of a plural food architecture. No single technology or practice constitutes the post-oppressive food system; what is required is a mosaic of approaches united by the shared commitment to care and solidarity as conditions of institutional admissibility. Josh Milburn's *Food, Justice, and Animals* (2023) offers the most systematic treatment of what such a plural, post-oppressive food system might look like within an animal-rights-respecting political order. The present argument converges with Milburn's in its rejection of industrial animal agriculture and its insistence that food justice requires institutional rather than merely individual transformation. Where Milburn operates within a liberal framework that sets limits on what the state can enforce and allows some forms of animal use under conditions of genuine respect, the argument here is more restrictive in what it permits – targeting the structural architecture of industrial food systems and treating their transformation as a demand of justice rather than a question of rights-compatible sourcing.

Cellular agriculture deserves particular attention within this plural framework because it is the site of the most significant political resistance and the most serious normative objections. Three objections require acknowledgement. The first concerns energy intensity: bioreactor production as currently configured is energy-intensive, and some lifecycle analyses suggest comparable or higher emissions than conventional poultry production under current energy mixes. This objection is real but not decisive: as El Wali et al. (2024) demonstrate, integration with renewable energy sources substantially alters the emissions profile, and the objection targets the current configuration of the technology rather than its potential.

The second concerns cell line sourcing: current cultivated meat production requires starter cells from living animals, and growth media have historically relied on foetal bovine serum. The chapter's own framework – which rejects trade in animal bodies and reproductive capacities – requires that serum-free and cruelty-free cell line technologies be treated not as optional refinements but as normative prerequisites for a just cellular agriculture. The industry's movement in this direction is the movement the framework demands, and public investment should accelerate rather than await it.

The third objection is the most structurally significant: the cultured meat industry is currently dominated by a small number of well-capitalised private firms, raising the prospect that cellular agriculture could reproduce the corporate concentration of industrial animal agriculture with the same consequences for small producers, labour, and democratic governance. This objection does not defeat the case for cellular agriculture but reshapes it: the justice of the technology depends entirely on the governance structures within which it is embedded. Cellular agriculture in the hands of large agribusiness reproduces industrial logic by technological means; cellular agriculture embedded within cooperative governance, public investment, and democratic accountability is a different institutional arrangement entirely. The argument for cellular agriculture is therefore inseparable from the argument for the governance architecture developed in the following section.

The transition to cell-based or cultured protein systems nonetheless faces significant obstacles. As chemical engineer David Humbird (2021) argued, bioprocess constraints – slow cell growth, high media costs, and limited bioreactor efficiency – make large-scale cultured meat production far from cost-competitive with conventional animal agriculture under current conditions.[6] Achieving industrial viability would require breakthroughs rather than incremental optimisation. Yet these limits are not purely technological. As El Wali et al. (2024) note, changes in public investment, regulation, and industrial priorities could substantially alter cost trajectories. Just as subsidies and externalised environmental harms have sustained fossil fuel dependence, animal agriculture has relied on structural economic privileges – feed and land-use subsidies, and the systematic exclusion of ecological, animal welfare, and social costs from market calculation. The economic future of cultured meat depends as much on collective political choices as on scientific progress.

Public funding for research and development, state procurement policies, and targeted subsidies can play decisive roles. Historical precedents in energy transition illustrate that cost barriers are not fixed: solar photovoltaics fell from one of the most expensive to one of the cheapest energy sources within two decades, primarily through sustained public investment and economies of scale (Kavlak et al., 2017). Similar dynamics could unfold in cellular agriculture if comparable political will were mobilised. The notion that cultured meat is too expensive to pursue functions less as a descriptive claim than as a political stance that accepts the status quo of animal oppression as the baseline of rational economy. Recognising this reveals the normative stakes: cost-efficiency arguments, when detached from justice considerations, risk reinforcing the very structures that render oppression profitable.

The political resistance to such alternatives underscores this dynamic. The recent banning of cultured meat sales and production in several United States jurisdictions illustrates that opposition is not merely technical but ideological. Texas Agriculture Commissioner Sid Miller celebrated the 2024 ban as a victory for Texas ranchers and consumers, invoking what he called a "God-given right" to know what is on one's plate and insisting that for millions of Texans, food must come from "a pasture, not a lab." Dismissing alternatives as contrary to "cowboy logic," he framed the ban as the defence of a "real, authentic meat industry" against synthetic competition (Texas Department of Agriculture, 2024). This rhetoric uses identity and authenticity to defend existing power structures in the food economy. The invocation of divine sanction, rural masculinity, and tradition situates animal exploitation within a moral imaginary that equates ethical reform with cultural betrayal – precisely the ideological construction that Carol Adams (1990) identified as central to the gendered politics of meat.

Such appeals to authentic meat are incongruent with the realities of contemporary animal agriculture. Industrial meat production depends on heavy technological mediation: routine use of antibiotics, antiparasitics, and other pharmaceuticals is necessary to maintain animals in conditions of confinement that would otherwise be fatal. In the United States, nearly 70 per cent of medically important antibiotics are used in livestock production, primarily to prevent disease in overcrowded environments rather than to treat illness (WHO, 2023). The natural pasture invoked by defenders of conventional meat functions as a nostalgic fiction, obscuring the fact that most meat is the product of a technologically engineered and pharmacologically sustained process. The cow in the feedlot is no more natural than the cells in the bioreactor.

The ideological contrast between pasture and laboratory inverts moral reality. The former relies on systems of confinement, premature death, and ecological degradation; the latter, at least in principle, makes nourishment possible without harm. To frame cultured meat as synthetic and conventional meat as authentic is to normalise violence and marginalise innovation. The rhetoric of authenticity disguises the moral cost of normalised cruelty as cultural virtue. The moral question is not whether new technologies depart from established practices but whether they reduce preventable suffering and enable fairer relations among different kinds of animals.

Taken together, the examples examined in this section show that it is already possible, in different registers, to decouple food provision from practices that negate animal self-development. Sanctuary-based agriculture achieves this by reclassifying animals from resources to residents; mixed agroecology does so by specifying ethical constraints on co-labouring; plant-based and cultured meat do so by removing animals from the production logic entirely. These are not utopian sketches but instantiated or modelled alternatives whose existence invalidates the claim that animal exploitation is a fixed precondition of feeding human populations. Their normative significance lies in falsifying the premise of necessity upon which the continued legitimacy of oppressive systems is commonly defended. If food can be produced without structural injustice, then oppression cannot be justified as an unavoidable cost – and the policy commitments that preserve industrial animal agriculture are therefore elective, subject to assessment and transformation. Justice is not a constraint to be traded off against efficiency after production is organised; it is a threshold condition for admission into the domain of what may count as a legitimate food system at all.

4 Toward a Post-Oppressive Food System

If food systems grounded in care and solidarity are to become more than marginal cases, their development must ultimately supplant rather than merely coexist alongside the prevailing paradigm of industrial animal agriculture. What is required is not incremental reform but a re-foundation of food production – one that abandons the organising logics of commodification, extractive value, and profit maximisation, and instead centres coexistence and relational responsibility across species lines. Such a transformation demands a conceptual and institutional rupture: the replacement of conditions that render exploitation normal and profitable with structures that make care, vulnerability, and self-development the non-negotiable terms of participation.

This vision entails a categorical rejection of animal commodification within the institutional architecture of industrialised food systems. A food system oriented by justice cannot coherently include slaughter, nor trade in animal bodies, reproductive capacities, or secretions as conditions of production. It cannot rely on markets that treat sentient lives as units of exchange, investment, or productivity. These are the minimal ethical requirements for coherence within a justice-oriented framework operating in the context this book addresses. A system that claims to embody justice cannot simultaneously reduce living beings to fungible assets, nor make their continued existence conditional upon economic yield or human preference. The precondition for participation in such a system is not utility but moral status, which demands inviolability, not harm management.

This claim is addressed primarily to the institutional architecture of contemporary industrialised food systems – the regime of confinement, commodification, and capital-intensive extraction that structures food production in industrialised societies. It does not purport to resolve the distinct questions raised by subsistence communities for whom animal use is not economically or ecologically optional, or by Indigenous food practices whose relationship to animal life is embedded in different historical, cultural, and ecological contexts. Those questions are real and require their own careful treatment. The argument here is narrower and more specific: that the industrial regime, which depends on none of the ecological or cultural justifications that might be advanced in those other contexts, cannot be rendered just through welfare reform or compensatory measures, and that its replacement is therefore a requirement of justice within the political communities this book addresses.

The shift toward such an order requires a fundamental reconfiguration of economic architecture. Rather than deriving value from the extraction of animal bodies or labour, revenue pathways must be anchored in plant-based production, ecosystem stewardship, and non-extractive multispecies relations. This may include ecological service economies, community-rooted horticulture and agroecology, food forests, and – where appropriate – cellular agriculture and fermentation-based protein production. The criterion is not novelty or efficiency but the institutionalisation of care and solidarity as the baseline condition for food production.

4.1 Temporal and Spatial Redesign

A transition toward care- and solidarity-based food systems requires a radical reconfiguration of the temporal and spatial organisations that structure agricultural practice. Contemporary animal agriculture organises land and time around extraction, acceleration, and turnover: animals are bred, grown, and slaughtered within compressed biological cycles; landscapes are engineered to maximise throughput and minimise the spatial footprint of sentient individuals. These are not neutral arrangements but the material expression of a political economy in which animal life is subordinated to productivity – from biological time to capital time (Moore, 2015) – and territory to commodity conversion rather than shared dwelling.

A clear illustration of this transmutation of biological time appears in Pachirat's (2011) *Every Twelve Seconds*. The book's title refers to the killing speed in the Nebraska beef plant where he worked undercover: the slaughter line was calibrated to kill one cow every twelve seconds – five per minute, hundreds per hour, thousands per day.[7] This tempo is dictated by the internal logics of industrial capitalism, whose requirements of efficiency and throughput are necessary to satisfy the dynamics of supply and demand.

The acceleration of slaughter must be understood alongside a parallel acceleration in animals' own biological development and a radical compression of their lifespans. A cow's natural lifespan is close to two decades, yet in industrial settings beef cattle are typically killed at around eighteen months of age. Male calves in the dairy industry – so-called "bob veal" – may be slaughtered just days, and rarely more than a few weeks, after birth. The temporal distortion is even more extreme in the case of chickens. Although their natural lifespan is approximately ten years, male chicks in the egg industry may live less than twenty-four hours, while broiler chickens bred for meat are slaughtered at roughly 40 days old.

A care-centred system inverts this organisation. Animal-inhabited landscapes would no longer be structured around yield extraction, reproductive control, or biological acceleration but around cohabitation, ecological regeneration, and multispecies security. Sanctuaries, habitat corridors, riparian buffers, and community-managed commons – supported through procurement rules, public investment, land-use regulation, and long-term stewardship frameworks – become foundational elements of this reorganisation. These are political infrastructures oriented toward coexistence, allowing animals to live, age, migrate, and form social bonds on their own terms while remaining part of shared social and ecological worlds.

Every governance instrument – subsidies, zoning codes, agricultural research funding, certification standards, liability rules – must answer to a single question: does this arrangement secure or foreclose the conditions of self-development? Together, such instruments constitute what might be called a new agrarian constitution, one in which food production no longer depends on premature death, deprivation, or displacement.

This reorientation also demands a reconceptualisation of time. A care-based farm does not operate on accelerated cycles of birth, growth, and slaughter; it follows lifetimes and seasons. Animals age, social groups persist, and planning incorporates long-term sanctuary obligations and intergenerational ecological regeneration. This stands in sharp contrast to industrial agriculture, which treats ageing as waste, mobility as inefficiency, and time itself as a variable to be compressed (Weis, 2013).

Space must likewise be understood as plural and ecologically textured. A care-centred landscape becomes a mosaic of habitats – refugia, shade belts, pollinator corridors, foraging strips, wetland margins, woodlots, and quiet zones – departing from the model of encapsulated zones for production, disposal, and consumption. Fencing, where present, serves safety and autonomy rather than confinement. Movement is no longer a cost but an expression of an ethic in which bodily agency and spatial freedom are central to multispecies self-development.

Under these conditions, ethical commitments are not appended to market systems but built into their architecture. Rather than adjusting harm within existing models, such a framework demands alternative value forms, temporalities, and land relations – ones in which humans and other animals participate as subjects of justice, not materials for production. Only then can food systems move beyond humane refinements of oppression toward genuinely emancipatory multispecies infrastructures of shared life.

4.2 Governance and Financial Infrastructure

A just food system requires governance and financial institutions that secure care as a public obligation rather than a private virtue. The organisation of food systems cannot be left to the aggregate choices of consumers – choices that are themselves largely shaped by market structures and corporate power. The state must assume responsibility for designing and regulating food infrastructures in ways that distribute benefits and burdens justly. When left to market logics alone, food systems inevitably determine who is included and who is rendered disposable. Democratic institutions, not consumer sovereignty, must serve as the guarantors of justice in this domain.

Decision-making cannot be oriented primarily toward the interests of industry owners or the artificially manufactured preferences of consumers whose desires are shaped by those same industries. Governance must instead recognise a plurality of affected parties and take seriously the ecological, social, and interspecies externalities that industrial agriculture routinely obscures. Care and solidarity must function as constitutional principles sustaining ecological integrity and respecting the lives and agencies of both human and nonhuman animals.

Within this framework, food-system governance becomes a multispecies practice of accountability. One institutional mechanism through which nonhuman interests could be represented within decision-making processes is the animal ombudsperson (see Magaña, 2022). Existing models remain embryonic and administratively peripheral: some jurisdictions grant animal ombudspersons limited procedural standing in administrative proceedings; others permit NGO representatives or enforcement actions; France grants animal protection associations *partie civile* status. These mechanisms are fragmented and largely confined to discrete violations rather than structural oversight of institutional design. Their significance lies less in what they currently achieve than in what they render institutionally imaginable: the systematic representation of nonhuman interests within the governance of food systems as a matter of justice rather than discretion. A caring food system would develop these embryonic forms into robust institutional arrangements capable of exercising ongoing scrutiny over the structural conditions – subsidies, procurement rules, land use, and liability frameworks – that determine how animals live and die within the food economy.

The distance between that vision and current reality is considerable. Few indicators reveal its structural pathology more clearly than the magnitude of global food waste. According to the 2024 Food Waste Index Report, nearly 30 per cent of the planet's agricultural land is effectively used to produce food that is never eaten (UNEP, 2024). Economically, more than one trillion US dollars in edible food is discarded annually – a scale of loss that reflects not isolated inefficiencies but deep systemic failures in production, distribution, and governance. Environmentally, food loss and waste account for an estimated 8–10 per cent of global greenhouse gas emissions, originating both in the supply chain and in the decomposition of discarded organic matter. These emissions compound the ecological pressures generated by land conversion for agriculture, which remains a principal driver of habitat destruction, biodiversity loss, and declining ecosystem resilience.

The moral dissonance becomes more acute when placed against global patterns of deprivation. While unprecedented quantities of food are wasted, approximately 783 million people experience chronic hunger each year, and around 150 million children under the age of five suffer stunting and impaired development due to persistent nutritional deficiencies. Understood through the framework of care developed in Chapter 6, food waste is not merely an efficiency failure or a distributive anomaly: it is a structural failure of attentiveness – a system organised around production and profit rather than around responsiveness to need and vulnerability, producing abundance while remaining constitutively indifferent to those it leaves hungry. What emerges is a global food regime structurally incapable of aligning production, distribution, and access with the needs of humans, nonhuman animals, and ecosystems, thereby reproducing deprivation and maldistribution even under conditions of material abundance.

Financially, just food systems depend on the internalisation of care costs rather than their displacement onto vulnerable beings or ecosystems. Public and cooperative funding mechanisms – for example, lifetime-care-based bonds (long-term public financial instruments tied to stewardship obligations rather than productivity metrics) for people, nonhuman animals, and ecosystems – would form the backbone of revenue structures, rewarding maintenance rather than extraction. This requires redirecting agricultural subsidies and research investments away from animal exploitation and toward practices that sustain multispecies well-being. Cooperative insurance schemes, care endowments, and community funds can provide long-term stability, ensuring that care obligations are insulated from market volatility.

Such arrangements reconceive profitability itself. Economic success is no longer measured by output volume or biological yield but by the durability and integrity of the relationships sustained within the system. Revenue derives from ecological regeneration, education, community engagement, and the provision of ecosystem services – each a form of value that extends rather than depletes shared life. The purpose of finance, in this model, is not to monetise care but to make it structurally possible.

4.3 Public Procurement and Regulatory Alignment

Public food procurement – the purchase of goods, services, and works by public authorities using public funds – is one of the most powerful and underutilised instruments available for scaling care- and solidarity-based food systems.[8] Because public institutions purchase food at scale, procurement decisions can determine not only what is bought – whether locally rooted, diverse, plant-forward, or nutritionally adequate – but also from whom and from which production systems (Swensson & Tartanac, 2020). When tied to justice-oriented standards, procurement can channel public funds toward small producers, community-based cooperatives, and farms committed to animal justice and ecological regeneration, reshaping both consumption and production patterns and embedding care and solidarity into supply chains rather than leaving them vulnerable to market pressures. Despite this transformative potential, public procurement remains widely underutilised, largely because regulatory frameworks rarely require institutions to align purchasing with obligations of ecological and animal justice.[9]

Regulatory frameworks should also evolve in tandem with these institutional and financial transformations. Zoning and land-use codes should explicitly recognise sanctuary-agriculture hybrids as legitimate and protected land uses. Such hybrids challenge the entrenched dichotomy between productive agricultural space and non-productive sanctuary space, instead treating animal-inhabited landscapes as sites of care, refuge, and ecological regeneration. This requires shifting away from legal regimes that categorise animals exclusively as property or production inputs – categories extensively critiqued in animal law and political theory (Francione, 2008; Deckha, 2021) – and toward some form of legal residency for animals (Donaldson & Kymlicka, 2011) – understood here in terms of inhabitance and standing rather than citizenship in the full political sense – that institutionalises their presence within territorial governance.

Liability regimes must also be restructured to reflect animals' capacities for agency, vulnerability, and self-development. Traditional tort systems treat harm to animals as a matter of property damage, erasing the relational dimensions of interspecies life. A justice-oriented alternative would allocate responsibility for harms in proportion to relational vulnerability and dependency, in line with the framework of structural and political responsibility developed in earlier chapters. Such frameworks distinguish between harms arising from shared ecological life and harms rooted in oppression, neglect, or exploitative relations, ensuring that the burden of risk does not fall disproportionately on the nonhuman animals whose vulnerability is most thoroughly produced by the systems they inhabit.

Subsidy structures must similarly be redesigned. Current agricultural subsidies overwhelmingly reinforce commodification by rewarding output volume, biological acceleration, and industrial efficiency (Kortleve et al., 2024; Rosenberg & Hayes, 2024; Weis, 2013). To prevent regression into these oppressive logics, public funding schemes should include explicit anti-commodification clauses that preclude support for production models premised on the systematic disposability of animal lives. Subsidies should instead prioritise regenerative practices, multispecies stewardship, and non-oppressive forms of food provisioning – consistent with proposals for post-growth transition pathways (Raworth, 2017; Jackson, 2017) and with the ethics of care and solidarity that underpin multispecies justice frameworks (Celermajer et al., 2021; Chrulew & Probyn-Rapsey, 2019). The goal is not merely to redirect funds but to transform the normative foundations of the agricultural state.

Taken together, these regulatory reforms recast land, liability, and subsidy regimes as levers for institutionalising animal justice. They move beyond efforts to mitigate harm within existing paradigms and instead embed care, responsibility, and justice as organising principles of food-system governance. By making care an enforceable condition of access to public resources, they invert the moral hierarchy that currently privileges efficiency and yield over justice and self-development. The state's fiscal and regulatory instruments thereby become constitutive of a new political economy of food – one in which public expenditure operates as an expression of collective moral responsibility.

4.4 Transparency, Scaling, and Justice Integration

Transparency and accountability are foundational to any food system governed by care and solidarity. This entails operating under open governance, with publicly accessible charters and transparent budgets and subsidies. Breaches of ethical standards should trigger independent investigations and result in those responsible being held accountable. Accountability here is not punitive but constitutive of a just food system, ensuring that the moral commitments of care remain visible, enforceable, and trusted by the communities they serve.

Scaling such systems should proceed not through volume expansion but through replication: distributed mosaics of farms and sanctuaries linked by shared principles, cooperative logistics, and mutual-aid insurance. The objective is not to build larger units of production but to multiply autonomous yet interdependent communities collectively composing a new agrarian order.

Boundary-setting is equally central. Clear compatibility rules must delineate what can and cannot be integrated into a justice-based food economy. Where financial pressures threaten these limits, public finance must close the gap, or the system must be redesigned to remain within its moral parameters.

The justice of care-based food systems must also extend to the human labour that sustains them. A multispecies ethic is incoherent if built on precarious or underpaid work. Cooperative ownership models, fair labour standards, and safe working conditions are not aspirational add-ons but structural imperatives. Transition pathways should support existing smallholders and workers, allowing movement from extractive models toward care-based ones without economic dispossession. Only by integrating human and nonhuman justice can these systems avoid reproducing the very hierarchies they seek to transcend. What emerges, when they succeed, is a genuinely emancipatory food economy grounded in reciprocity and shared self-development.

5 Conclusion

This chapter has elaborated the normative foundations and institutional design of a food system grounded in interspecies care and justice. Taken together, these elements sketch the structural framework of an order built on care and solidarity rather than extraction and disposability – one that translates the moral claim that animals are subjects of justice into a concrete political and economic configuration, organised around temporal and spatial redesign, democratic governance and finance, regulatory and procurement alignment, and robust mechanisms of accountability and scaling. What emerges is not a utopian enclave but a viable alternative, anchored in existing counter-regimes and in the demonstrable possibility of nourishing populations without making oppression a precondition of production.

This shift clarifies what is at stake in the politics of transition. Industrial animal agriculture is not simply a flawed or excessively harmful sector awaiting refinement; it is a regime whose organising logic depends on confinement, premature death, and ecological displacement, and whose legitimacy therefore cannot be restored through compensatory reform. Once the material possibility of post-oppressive provisioning is established – whether through plant-based systems, cellular agriculture, or care-centred agroecologies – continued reliance on commodified sentience appears not as a necessity but as a political choice. The relevant question is then not how much harm can be mitigated within the existing paradigm but why a system predicated on structural injustice should remain eligible as a legitimate option.

Replacing industrial animal agriculture is therefore not only an ethical imperative but a democratic task. It requires shifting the locus of responsibility from individual consumption to public institutions that currently make exploitation profitable and care exceptional. It also demands institutional forms capable of representing nonhuman interests in the design and oversight of food governance, so that animals' vulnerability is treated not as a background cost but as a matter of public accountability.

A caring food system is best understood as a constitutional achievement: a redefinition of admissible economic relations in light of shared vulnerability and interdependence. Its guiding commitment is simple – bodily integrity and developmental capacities must not be treated as disposable for the sake of efficiency, preference, or profit. Only when care functions as a condition of institutional admissibility, rather than an afterthought of regulation or consumer virtue, can the food economy become a site of animal justice. Under such conditions, the boundaries of the polity extend its obligations beyond the human, so that coexistence, reciprocity, and mutual self-development become the polity's organising principles.

Notes

1 The claim that industrial food systems privilege efficiency and profit must itself be questioned. According to a recent report by the Food System Economics Commission, while generating approximately $10 trillion annually, the food system produces an estimated $15 trillion in hidden environmental and social costs – including worsening health outcomes, lost productivity, ecosystem degradation, and significant burdens on public budgets (Ruggeri Laderchi et al., 2024). The system does not merely distribute efficiency gains unjustly; it generates more harm than value. What presents itself as economic rationality is, by its own metric, a failure.
2 For examples of this perspective, see Brambell (1965); Grandin and Johnson (2005); Fraser (2008).
3 The extraordinary extension of public subsidies for animal agriculture exemplifies this clearly. See Kortleve et al. (2024).
4 Perfecto et al. (2019) also critique conservation groups for supporting the land-sparing strategy, which promotes intensifying agriculture on some land to leave other areas untouched. Instead, they advocate for conservationists to support land reform and food sovereignty efforts, which can create the conditions necessary for more small, diverse farms – landscapes that serve as valuable habitats for a wide range of species.
5 The Commission members emphasise that unsustainable and unhealthy food systems pose a serious threat to both human health and planetary well-being. While hundreds of millions suffer from hunger, even more consume diets that fuel chronic disease and premature death. At the same time, current methods of food production are a major driver of environmental degradation. With global diets unchanged and the population projected to reach 10 billion by 2050, the associated health risks and ecological pressures – including greenhouse gas emissions, nutrient runoff, biodiversity loss, and intensive use of water and land – are expected to escalate significantly.
6 Recent industry analysis suggests that second-generation cultivated meat companies have already made significant progress in overcoming the cost barriers identified by Humbird (2021). See Lever VC (2025).
7 Early in 2025, the Trump administration announced new regulatory directives intended to ease compliance burdens on the United States pork and poultry industries, thereby permitting higher line speeds and greater operational efficiency. Prior to these changes, most poultry plants in the United States operated at approximately two birds per second, while pork processing facilities were authorised to slaughter up to 1,106 animals per hour. See United States Department of Agriculture (2025): www.usda.gov/about-usda/news/press-releases/2025/03/17/secretary-rollins-takes-action-streamline-us-pork-and-poultry-processing
8 For a report on the capacity of food procurement to contribute to the right to food by a United Nations expert, see De Schutter (2014).
9 For an example of ecological procurement with more focus on sustainability than interspecies justice, see Erdmenger (2003).

References

Adams, C. J. (1990). *The sexual politics of meat: A feminist-vegetarian critical theory*. Continuum.

Akter, S. et al. (2025). Integrated farming system – A means of improving farmers' well-being in the wetland ecosystem of Bangladesh. *Farming System*, 3, 100127. doi:10.1016/j.farsys.2024.100127.

Altieri, M. A. (1995). *Agroecology: The science of sustainable agriculture* (2nd ed.). Taylor & Francis.

Altieri, M. A., & Nicholls, C. (2005). *Agroecology and the search for a truly sustainable agriculture*. United Nations Environment Programme.

Appiah, K. A. (2010, January 8). What will future generations condemn us for? *The Washington Post.* www.washingtonpost.com/wp-dyn/content/article/2010/09/24/AR2010092404113.html.
Brambell, F. W. R. (1965). *Report of the technical committee to enquire into the welfare of animals kept under intensive livestock husbandry systems.* Her Majesty's Stationery Office.
Brown, W. (2015). *Undoing the demos: Neoliberalism's stealth revolution.* Zone Books.
Celermajer, D. et al. (2021). Multispecies justice: Theories, challenges and a research agenda for environmental politics. *Environmental Politics*, 30 (1–2), 119–140.
Chiles, R. M. (2023). The agrarian ethic and the regenerative promise. *Agriculture and Human Values*, 40, 45–61. doi:10.1007/s10460-022-10382-9.
Chrulew, M., & Probyn-Rapsey, F. (Eds.). (2019). *Animal death.* Sydney University Press.
Clapp, J. (2020). *Food* (3rd ed.). Polity Press.
Clark, M. et al. (2021). Multiple health and environmental impacts of foods. *Proceedings of the National Academy of Sciences*, 118 (9), e2012389118. doi:10.1073/pnas.2012389118.
De Schutter, O. (2014). *The power of procurement: Public purchasing in the service of realizing the right to food.* United Nations Special Rapporteur on the Right to Food.
de la Bellacasa, M. P. (2017). *Matters of care: Speculative ethics in more than human worlds.* University of Minnesota Press.
Deckha, M. (2021). *Animals as legal beings: Contesting anthropocentric legal orders.* University of Toronto Press.
Donaldson, S., & Kymlicka, W. (2011). *Zoopolis: A political theory of animal rights.* Oxford University Press.
El Wali, M. et al. (2024). Transition to cellular agriculture reduces agriculture land use and greenhouse gas emissions but increases demand for critical materials. *Communications Earth & Environment*, 5, 61.
Erdmenger, C. (Ed.). (2003). *Buying into the environment: Experiences, opportunities and potential for eco-procurement.* Routledge. doi:10.4324/9781351281409.
FAO. (2019). The 10 elements of agroecology: Guiding the transition to sustainable food and agricultural systems. https://openknowledge.fao.org/server/api/core/bitstreams/3d7778b3-8fba-4a32-8d13-f21dd5ef31cf/content.
Francione, G. L. (2008). *Animals as persons: Essays on the abolition of animal exploitation.* Columbia University Press.
Fraser, D. (2008). *Understanding animal welfare: The science in its cultural context.* Wiley-Blackwell.
Ghaznavi, S. et al. (2025). Sustainable synergies of rice–fish–duck symbiosis: A life cycle assessment in Guilan Province of Iran. *Results in Engineering*, 28, 107226.
Giraud, E. H. (2021). *What comes after entanglement? Activism, anthropocentrism, and an ethics of exclusion.* Duke University Press.
Grandin, T., & Johnson, C. (2005). *Animals in Translation: Using the Mysteries of Autism to Decode Animal Behavior.* Harcourt, Brace, Jovanovich.
Harari, Y. N. (2015, September 25). Industrial farming is one of the worst crimes in history. *The Guardian.* www.theguardian.com/books/2015/sep/25/industrial-farming-one-of-worst-crimes-history-ethical-question.
Holt-Giménez, E. (2017). *A foodie's guide to capitalism: Understanding the political economy of what we eat.* Monthly Review Press.
Howard, P. H. (2016). *Concentration and power in the food system: Who controls what we eat?*Bloomsbury Academic.

Humbird, D. (2021). Scale-up economics for cultured meat. *Biotechnology and Bioengineering*, 118 (8), 3239–3250.

IPCC. (2019). *Climate change and land: An IPCC special report on climate change, desertification, land degradation, sustainable land management, food security, and greenhouse gas fluxes in terrestrial ecosystems*. Intergovernmental Panel on Climate Change.

Jackson, T. (2017). *Prosperity without growth: Foundations for the economy of tomorrow* (2nd ed.). Routledge.

Kavlak, G., McNerney, J., & Trancik, J. (2017). Evaluating the Causes of Cost Reduction in Photovoltaic Modules. *Sustainable Technology eJournal*. doi:10.2139/ssrn.2891516.

Kortleve, A. J. et al. (2024). Over 80% of the European Union's Common Agricultural Policy supports emissions-intensive animal products. *Nature Food*, 5, 288–292.

Lever VC. (2025, April). A second generation of cultivated meat companies breaks through projected cost barriers. www.levervc.com/wp-content/uploads/2025/04/Lever-VC-A-Second-Generation-of-Cultivated-Meat-Companies-Breaks-Through-Projected-Cost-Barriers-4.pdf.

Magaña, P. (2022). The political representation of nonhuman animals. *Social Theory and Practice*, 48 (4), 665–690.

Milburn, J. (2023). *Food, justice, and animals: Feeding the world respectfully*. Oxford University Press.

Moore, J. W. (2015). *Capitalism in the web of life: Ecology and the accumulation of capital*. Verso.

Nussbaum, M. C. (2006). *Frontiers of justice: Disability, nationality, species membership*. Harvard University Press.

Pachirat, T. (2011). *Every twelve seconds: Industrialized slaughter and the politics of sight*. Yale University Press.

Perfecto, I., & Vandermeer, J. (2015). *Coffee agroecology: A new approach to understanding agricultural biodiversity, ecosystem services and sustainable development*. Routledge.

Perfecto, I. et al. (2019). *Nature's matrix: Linking agriculture, biodiversity conservation and food sovereignty* (2nd ed.). Routledge.

Philpott, S. M., & Bichier, P. (2012). Effects of shade tree removal on birds in coffee agroecosystems in Chiapas, Mexico. *Agriculture, Ecosystems & Environment*, 149, 171–180.

Raworth, K. (2017). *Doughnut economics: Seven ways to think like a 21st-century economist*. Chelsea Green Publishing.

Rosenberg, B., & Hayes, J. (2024, October 28). USDA livestock subsidies top $72B. *Environmental Working Group*. www.ewg.org/news-insights/news/2024/10/usda-livestock-subsidies-top-59-billion.

Ruggeri Laderchi, C. et al. (2024). *The economics of the food system transformation: Global policy report*. Food System Economics Commission. https://foodsystemeconomics.org/wp-content/uploads/FSEC-Global_Policy_Report.pdf.

Sciortino, C. et al. (2025). Greenwashing in the agri-food industry: A discussion around EU policies and the Italian context. *Food and Humanity*, 4, 100528.

Shamir, R. (2008). The age of responsibilization: On market-embedded morality. *Economy and Society*, 37 (1), 1–19.

Siebrecht, N. (2020). Sustainable agriculture and its implementation gap – Overcoming obstacles to implementation. *Sustainability*, 12 (9), 3853. doi:10.3390/su12093853.

Singer, P. (2009). *Animal liberation: The definitive classic of the animal movement* (updated ed.). Harper Perennial.

Steinfeld, H. et al. (2006). *Livestock's long shadow: Environmental issues and options*. FAO.

Swensson, L., & Tartanac, F. (2020). Public food procurement for sustainable diets and food systems: The role of the regulatory framework. *Global Food Security*, 25, 100366.

Texas Department of Agriculture. (2024, May 29). *Commissioner Miller praises Texas Legislature for putting a fork in lab-grown meat*. https://texasagriculture.gov.

Twine, R. (2010). *Animals as biotechnology: Ethics, sustainability and the politics of animal enhancement*. Earthscan.

Twine, R. (2021). Emissions from animal agriculture – 16.5% is the new minimum figure. *Sustainability*, 13 (11), 6276.

UNEP. (2024). *Food waste index report 2024*. United Nations Environment Programme.

Weis, T. (2013). *The ecological hoofprint: The global burden of industrial livestock*. Zed Books.

WHO. (2023). *Critically important antimicrobials for human medicine* (7th rev.). World Health Organization.

Willett, W. et al. (2019). Food in the Anthropocene: The EAT–Lancet Commission on healthy diets from sustainable food systems. *The Lancet*, 393 (10170), 447–492.

Wu, D., & Liu, J. (2024). Promoting sustainable agroforestry development: A systematic literature review on the rice–fish–duck–forest system. *Environment, Development and Sustainability*. doi:10.1007/s10668-024-05601-6.

Young, I. M. (2000). *Inclusion and democracy*. Oxford University Press.

Zhang, Y., et al. (2023). Review of rice–fish–duck symbiosis system in China – One of the globally important ingenious agricultural heritage systems (GIAHS). *Sustainability*, 15 (3), 1910. doi:10.3390/su15031910.

9

BUILDING CARE

Urban Space and Animal Justice

1 Introduction

The preceding chapters examined how the values of care and solidarity might be embedded, first, in constitutional design and, second, in the organisation of the food system. Yet political principles are not fully realised at the level of foundational norms or sectoral regulation alone. If oppression is structural, it must also be addressed through the material arrangements that shape everyday life. Care and solidarity must therefore also be inscribed in the architecture and urban design of the polity itself. In this chapter, I turn to the city as the site where constitutional commitments and material systems converge, asking how an interspecies politics might take form in the organisation of urban life.

More precisely, I ask what a polity grounded in care and solidarity would require in its spatial design. The aim is not to provide a comprehensive blueprint but to identify the structural features of an interspecies polity that recognises animals as subjects of justice and secures the material conditions necessary for self-development across species. Among the most durable mediations of political values are built environments. They organise patterns of interaction, vulnerability, and exposure to harm. They determine who counts, whose presence is legitimate, and whose needs must be accommodated.

Political principles become operative when they are incorporated into arrangements such as zoning regimes, transport systems, housing policies, public health infrastructures, and environmental regulation. Urban space is not normatively neutral; it reflects and reproduces background assumptions about whose lives matter and whose do not. This perspective aligns with Iris Marion Young's structural account of justice, according to which institutions embody patterns of power and privilege that are always also spatially organised,

DOI: 10.4324/9781003434856-11

shaping the possibilities for self-development through the material arrangements of collective life (Young 1990, 2000).

Urban design distributes risks, resources, and opportunities unevenly, not only among human groups but across species. An interspecies politics of care must therefore attend to how these systems are designed and whom they are designed for.

If animals are to be recognised as inhabitants of a shared political space, their interests cannot remain external to urban planning, regulatory regimes, and public infrastructure. The question, then, is not merely whether animals are protected by law, but whether urban institutions are structured to welcome their presence, reduce their vulnerability, and enable forms of self-development for beings of various species. The remainder of this chapter examines what such structuring would require.

2 The Architecture of the City

Cities are organised around background conceptions of value. They materialise political principles and judgements about whose presence matters, whose activities are legitimate, and which forms of dependence must be accommodated. Urban design – through zoning, transport systems, housing patterns, and environmental regulations – reflects and stabilises hierarchies of worth. The city is thus structured for certain beings, capacities, and modes of life, while rendering others marginal, burdensome, or disposable. Justice, therefore, is not exhausted by formal rights or legislative declarations. In Iris Marion Young's language, it is embedded in the structures of collective life that shape opportunities for self-development (Young, 1990). The urban landscape is one such structure.

Modern cities have long been organised according to logics that include some forms of life while excluding others. A well-known example is what has come to be called Haussmannisation, referring to the large-scale redesign of Paris under Georges-Eugène Haussmann in the nineteenth century (Jordan, 1995). Haussmann's transformation of Paris illustrates how spatial design can institutionalise power. The new boulevards improved circulation and sanitation, but they simultaneously displaced working-class populations and facilitated surveillance and military control. Urban restructuring thus redistributed security, visibility, and access in ways that advantaged some groups and burdened others.

Urban design is never merely technical. Decisions about street width, housing density, green space, transport systems, and waste management shape patterns of vulnerability and exposure. They determine who can remain in place, who must relocate, and under what conditions collective life unfolds. Contemporary mega-projects that disregard existing socio-spatial relations (Gellert & Lynch, 2003) continue this pattern, privileging growth, investment, and efficiency over stability and continuity.

This spatial organisation can be read through Iris Marion Young's account of structural injustice. The city distributes the five faces of oppression she identified not only through laws and labour markets but through infrastructure itself. Exploitation is facilitated when housing markets and transport systems extract labour while relegating workers to peripheral zones. Marginalisation occurs when zoning regimes and public-space regulations displace the unhoused, migrants, or informal workers. Powerlessness is reproduced when decision-making about urban development is insulated from those most affected by it. Cultural imperialism is embedded in architectural norms and aesthetic standards that treat some forms of life as normal and others as deviant, disorderly, or alien. Violence is structured into policing practices and environmental exposure, where some populations are made systematically more vulnerable to harm.

If justice concerns the institutional conditions of self-development, then the built environment is one of its primary mediators. Urban space shapes mobility, association, safety, access to resources, and risk exposure. It conditions who can exercise capacities freely and whose development is persistently constrained.

These dynamics extend beyond intra-human inequality. The very institutional mechanisms that marginalise certain human groups also organise the city along species lines. Zoning regimes, nuisance ordinances, sanitation codes, and wildlife management policies determine which nonhuman animals may inhabit urban space and on what terms. Some are incorporated as companions or as managed wildlife; others are classified as pests and subjected to removal or extermination. As argued in Chapter 2, this constitutes a species-structured form of marginalisation and violence, embedded not in episodic acts of cruelty but in the ordinary administrative and spatial ordering of urban life.

A politics of care and solidarity must therefore confront urban structure itself. If oppression is spatially organised, care cannot remain merely attitudinal or interpersonal. It must be institutionalised in planning, regulation, and infrastructure, reshaping the city's material and legal arrangements so that they enable rather than systematically foreclose multispecies self-development.

3 Anthropocentric Urbanism

In most contemporary cities, the hierarchies that structure human relations – between rich and poor, citizen and migrant, owner and labourer – are mirrored in the city's treatment of nonhuman life. Urban design systematically prioritises human mobility, convenience, and consumption, often at the expense of other species. Glass façades and illuminated high-rises disorient and kill migratory birds; traffic noise and artificial light disrupt navigation and breeding patterns (van Doren et al., 2017; (van Doren 2021; Longcore & Rich, 2004); storm drains, fences, and road networks fragment habitats and obstruct animal movement (Zellmer & Goto, 2022). Infrastructure built without regard for nonhuman life kills animals as a matter of routine – not through malice but through indifference encoded in design.

These effects are not incidental externalities. They reflect the organisation of the city as a space designed primarily for human ends – and that design extends beyond the treatment of wild animals to the systematic concealment of how domesticated animals are used. Animals raised for food are displaced to industrial facilities at the urban periphery, spatially removed from everyday visibility.[1] This segregation sustains what earlier chapters described as privileged irresponsibility: the capacity to benefit from systematic exploitation while remaining insulated from its conditions and consequences. Urban residents consume animal products while the infrastructures of confinement, slaughter, and waste are kept physically and symbolically distant.

The result is a patterned exclusion of nonhuman animals from the spatial community. Urban modernity treats animals either as controlled companions, managed wildlife, or nuisances to be eliminated. As Jennifer Wolch has argued across a series of influential essays, the modern city is structured through the exile of the animal – the removal of nonhuman life from the recognised boundaries of the polis (Wolch, 1998, 2002). What appears as an environmental side-effect is therefore better understood as a constitutive feature of urban design grounded in an anthropocentric conception of political space – one that impoverishes the city itself, stripping it of what Wolch calls the *anima urbis*, the breath and life embodied in its nonhuman as well as human inhabitants. In opposition to this logic, I follow Wolch in developing an alternative urban vision in which animals are reintegrated into city life as recognised cohabitants rather than managed populations or nuisances.

The development of the modern city can thus be understood as a progressive regulation of space in which forms of life deemed disorderly or disruptive are contained, relocated, or excluded. Animals that do not fit established categories – neither fully domesticated nor fully wild – occupy a liminal status within this framework. Their presence is tolerated only insofar as it remains controlled or marginal. Rats, pigeons, foxes, crows, stray dogs, and others persist in the residual spaces of urban design, adapting to infrastructures built without regard for their interests. Their classification as pests reflects not merely ecological concern but an evaluative judgement about which forms of life are legitimate inhabitants of the city.

Urban political ecology has emphasised that such exclusions are structured through spatial classification – through the regulation of what counts as "in place" and "out of place" (Cresswell, 1996, 2004). Designations such as "clean," "safe," or "developed" areas do more than describe physical conditions; they encode normative assumptions about order and belonging. Spaces associated with waste, decay, or animal presence are treated as zones requiring management or removal. The spatial segregation of slaughterhouses to industrial peripheries, the relocation of unhoused populations to shelters, and the removal of carcasses to rendering facilities are not incidental arrangements.

They produce moral distance by separating sites of dependency, vulnerability, and death from everyday visibility. The contemporary city thereby sustains an image of autonomy and control, obscuring the ecological and bodily dependencies on which it rests.

These exclusions reflect a long-standing association between political recognition and capacities such as speech, rational agency, and public participation, criteria that have historically been treated as threshold conditions for inclusion in political life. Nonhuman animals, presumed not to meet these standards, are positioned outside the recognised boundaries of political concern. The city gives institutional and spatial form to this exclusion. Its infrastructures of hygiene, order, and circulation are organised on the assumption that legitimate occupants are, at least primarily, humans. Even where animals are accommodated – pets in apartments, ornamental fish in fountains, or horses in police units – the terms of inclusion typically reinforce human authority. Animals are incorporated as companions, symbols, or instruments of security, not as subjects whose presence generates independent claims on urban space.

Approaching the city through care and solidarity requires re-examining these background assumptions. Care, as developed in Chapter 6, begins with attentiveness and responsiveness to vulnerability. In the urban context, this entails recognising animals not merely as managed populations or environmental variables but as cohabitants whose vulnerability and interests must be considered in planning and governance. Solidarity, in turn, requires more than accommodation or tolerance; it calls for institutional arrangements that enable forms of interspecies co-existence consistent with the self-development of all those who inhabit a common political space.

The structural conditions faced by different categories of animal presence differ significantly, and a politics of care must be responsive to these differences rather than treating all animal presence through a single lens. Wild animals are affected primarily through habitat loss and fragmentation; liminal animals (Donaldson & Kymlicka, 2011) face routine classification as infestations despite their ecological integration into urban life; domesticated animals used in food production are subject to the further injustice of systematic invisibility – embedded in urban economies while their conditions of production remain hidden; companion animals, by contrast, are highly visible but often reduced to instruments of human preference, status, or commerce, their welfare contingent on the whims of ownership rather than secured as a matter of justice.

4 Designing Care

Designing for care is not a matter of symbolic inclusion or gestures (Tronto, 1993). It requires reorganising the material systems that sustain collective life – transport, housing, waste management, food distribution, green space – so that they enable the self-development of multiple species. To design with care is to acknowledge that the presence of others generates normatively binding claims on how space is organised. It entails anticipatory planning such as building corridors rather than barriers, foreseeing harm rather than merely responding to disaster, and incorporating nonhuman interests into zoning regimes, environmental regulation, and municipal governance. A just multispecies city understands such planning as the continual adjustment of environments to conditions of shared vulnerability, treating this task not as a secondary concern but as a central organising function of urban political life.

In contrast to a Haussmann-like urban design, a city organised around care would resist the segregation that characterises conventional urban design. Rather than enclosing and separating, it would favour ecological and social continuity – interconnected parks, waterways, and green corridors that allow movement across human and nonhuman domains (Beatley, 2011; Lefosse et al., 2025). Urban environments must also accommodate changing needs across species: flexible zoning, mixed-use spaces, and materials that reduce rather than intensify harm are expressions of a commitment to shared inhabitation. And rather than continually extending urban frontiers, a caring city would redirect energy toward maintenance and restoration, addressing existing ecological fragmentation, environmental degradation, and infrastructural inequities rather than producing new ones.

Caring design is already operative in specific planning practices. Wildlife corridors and ecological overpasses reconnect fragmented habitats and allow animals to traverse highways and railways with reduced risk. Long-term research in Banff National Park, in Canada, demonstrates that well-designed crossing structures significantly facilitate large mammal movement across roads, with effectiveness varying by species, structural attributes, and levels of human activity near the crossings (Clevenger & Waltho, 2000, 2005). A comparable European example is the Zagreb–Rijeka highway through Gorski kotar in Croatia, designed with viaducts, tunnels, and one specifically constructed green bridge 100 metres wide, where monitoring confirmed that brown bears, grey wolves, and Eurasian lynx all used the crossing structures regularly and the highway was found not to function as a barrier to large carnivore movement (Huber et al., 2018). Urban green infrastructure initiatives in cities such as Amsterdam and Melbourne similarly integrate waterways, tree canopies, and vegetated rooftops into broader planning strategies, simultaneously reducing heat-island effects and expanding habitats for birds, insects, and small mammals. These examples illustrate how institutionalised planning choices can operationalise care through material design.

Caring design expands the range of recognised users of infrastructure and the range of individuals whose interests its existence is understood to affect. Rather than treating built systems solely as instruments for human mobility and consumption, it structures them to accommodate multispecies presence. Infrastructure need not function through rigid separation: a stormwater canal can be engineered to serve as a wetland habitat; a bridge can incorporate wildlife passage alongside human transit; a public square can be designed with shade, water access, and planting schemes that support birds and insects as well as people. In each case, design choices reflect underlying evaluative commitments that determine whose needs are anticipated in advance and whose are treated as incidental. To design with care is to institutionalise attentiveness and incorporate the foreseeable interests of others into the planning criteria that guide urban development.

The ethical significance of these interventions lies in their redefinition of planning priorities. Conventional urban metrics often assess success in terms of speed, efficiency, property value, and aesthetic order – criteria centred on particular forms of human use. A politics of care instead evaluates urban design in terms of shared liveability: the extent to which the built environment sustains multiple forms of life capable of self-development without systematically subordinating some to the convenience of others. It is a spatial expression of solidarity insofar as it recognises that well-being is relational and structurally interdependent. Such considerations must be integrated across scales of governance and design, from building façades and public lighting to transport systems, metropolitan planning, and the legal protection of ecological corridors, embedding care into regulatory standards rather than treating it as discretionary goodwill.

Extending this orientation to urban governance requires both cultural change and institutional design. Planning procedures could formally incorporate expertise from animal welfare organisations, ecologists, and affected local communities. Environmental impact assessments could be supplemented by multispecies justice assessments evaluating foreseeable effects on both human and nonhuman inhabitants. Zoning regulations and infrastructure standards could mandate provisions for ecological connectivity, wildlife safety, and habitat preservation – understood not as ends in themselves but as conditions for the self-development of the sentient animals who depend on them. Animal care would become a condition of legitimate urban development rather than a discretionary concern.

Finally, a caring urban aesthetic would treat the presence of nonhuman animal life as compatible with, and even indicative of, successful design. The measure of urban quality would not be the elimination of all irregularity but the capacity of space to sustain diverse forms of life without systemic oppression. The caring city is not defined by pristine uniformity but by ecological functionality and institutional responsiveness.

5 Environmental Justice and the Urban Metabolism

If care in design concerns the spatial organisation of co-existence, then care as environmental justice concerns the metabolism of the city – the regulated circulation of air, water, energy, food, and waste that sustains urban life (Kennedy et al., 2007). Cities do not merely occupy space; they transform and redistribute environmental goods and harms. Clean air and polluted air, access to green space and exposure to toxins, shade and heat, silence and noise are distributed through infrastructural and regulatory decisions. These distributions track and reinforce existing hierarchies of class, race, and species – patterns of advantage and exposure that the organisation of urban life systematically reproduces.

Thus, an interspecies politics of care requires expanding environmental justice beyond its conventional anthropocentric frame. It must examine how the benefits and burdens generated by urban metabolism are allocated across all inhabitants of the city. To treat animals as subjects of justice is to recognise that they too are positioned within these metabolic systems, exposed to pollution, heat, habitat degradation, and resource scarcity. In this account, environmental justice becomes a multispecies question.[2]

The traditional framework of environmental justice emerged in response to the unequal distribution of environmental burdens among human communities. It demonstrated how low-income and racialised populations are disproportionately exposed to pollution, toxic waste, and degraded environments (Bullard, 1990; Wenz, 1995; Shrader-Frechette, 2002). The central insight – that environmental harms track lines of social power – extends to the animal context. The same infrastructural and regulatory decisions that concentrate pollutants in marginalised human neighbourhoods also affect nonhuman animal populations inhabiting those spaces. Animals living near industrial zones, highways, or contaminated waterways are exposed to the same airborne toxins, heavy metals, and habitat degradation as the human communities alongside them. The difference is not exposure but recognition: their injuries rarely register within environmental monitoring systems or public deliberation.

Urban environmental injustice is therefore dual in structure. It distributes harms unequally among human populations while simultaneously excluding nonhuman victims from frameworks of accountability. Consider industrial animal production facilities, whose externalities propagate through urban and peri-urban ecosystems even when the facilities themselves are sited at distances from city centres. They externalise methane emissions, nutrient runoff, and waste into surrounding ecosystems, imposing ecological burdens on neighbouring communities and wildlife alike. Runoff contaminates waterways, affecting fish and amphibians; air pollution harms birds and companion animals; yet these impacts seldom figure in urban planning metrics or regulatory assessments. The metabolic processes that sustain the city thus generate multispecies costs that remain institutionally unacknowledged.

An interspecies approach reframes these dynamics as failures of solidarity. Solidarity, in this context, involves recognising shared vulnerability within the city's metabolic systems and accepting collective responsibility for mitigating their unequal effects. Pollution, waste, and resource extraction become not merely technical challenges but the expression of structured relationships in which some populations – human and nonhuman – bear unequal burdens.

David Schlosberg (2007) has rightly argued that environmental justice encompasses not only distribution but also recognition and participation.

Extending this framework across species requires expanding the scope of institutional consideration. While nonhuman animals cannot participate directly in political deliberation, their interests can be represented through structured mechanisms – ecological ombudspersons, biodiversity councils, or advisory bodies that integrate ecological science and ethical analysis into urban planning. Such institutions would not anthropomorphise nonhuman agency but would ensure that foreseeable ecological impacts are incorporated into decision-making procedures, approximating what Nancy Fraser (2008) describes as participatory parity – the condition under which those affected by social arrangements are included within the structures that govern them – through designated representation rather than direct participation.[3]

A concept that runs through contemporary urban policy – and that bears directly on how metabolic burdens are distributed – is resilience (Yamagata & Sharify, 2018; Masnavi et al., 2019). In its dominant usage, resilience denotes the capacity of communities or ecosystems to absorb disturbance without significant transformation. While adaptive capacity is important, an exclusive focus on resilience can shift attention away from the structural causes of harm and toward the endurance of those most affected. A justice-oriented account would instead emphasise transformative resilience: reducing exposure through redistribution and redesign rather than normalising unequal vulnerability. Rather than promoting adaptation to adverse conditions, the objective is to alter the conditions that produce them.

Transformative resilience, so understood, demands the redistribution of metabolic benefits and burdens – green space, clean air, noise, light, waste exposure – not as a technical optimisation problem but as a matter of justice owed to all those, human and nonhuman, whose well-being depends on how those flows are organised. The metabolic systems of the city – energy, water, and waste – provide concrete domains for institutionalising care. In each, care takes the form of maintenance and restoration: aligning urban production and consumption with the ecological systems on which they depend and the sentient animals who inhabit them.

Environmental justice must also address the uneven geography of harm within the city itself. Industrial corridors, waste facilities, and major transport routes frequently intersect with marginalised human neighbourhoods and with nonhuman animals occupying the same spaces. Birds nest in bridge structures above highways; stray dogs forage near disposal sites; rodents inhabit sewer systems exposed to chemical residue. These populations are systematically exposed to elevated environmental risk, and their conditions are structurally linked to patterns of urban privilege elsewhere. To design for solidarity is to render these relations visible within planning processes and to treat them as matters of institutional responsibility rather than incidental side effects.

Achieving this requires integrating environmental ethics into the core decision-making structures of urban governance – not only within municipal boundaries but beyond them. Cities externalise ecological costs to distant regions through energy consumption, material sourcing, and waste disposal, displacing the burdens of urban life onto ecosystems and sentient animals far removed from the spaces of urban comfort. An interspecies politics of care must therefore extend beyond the administrative city, acknowledging that urban well-being is sustained by ecological relations that stretch across extraction sites, deforestation frontiers, and climate-affected habitats. Justice requires that these entanglements be made visible within planning processes and treated as matters of institutional responsibility rather than externalities.

When institutionalised within urban metabolism, care and solidarity organise governance around precaution, responsibility, and restoration – preventing foreseeable harm, allocating burdens in proportion to benefits, and addressing accumulated ecological damage through regeneration and redistributive policy.

The caring city is thereby situated within a broader project of equitable liveability. It conceives the city not as a self-contained arena of human exchange but as an open metabolic system whose operations generate obligations across species and temporal scales. To care for the city, in this sense, is to structure its flows of energy, materials, waste, and risk so that they do not systematically externalise harm onto the vulnerable. The following section turns from these material dimensions to the civic and cultural conditions of justice, examining how public space and shared practice can cultivate the interspecies civic ethos that a multispecies polity requires.

6 Civic Dispositions and the Education of Perception

The preceding sections have examined the structural and material dimensions of the caring city – its spatial organisation and its environmental metabolism. Yet just institutions also require corresponding dispositions which, as argued in Chapter 6, must be grounded in care and solidarity. A multispecies polity depends not only on formal arrangements but on a civic orientation that recognises interdependence as a basic condition of political life. Cities, by virtue of their density and diversity, are particularly significant in this regard. Through their design and everyday practices, they shape how citizens encounter difference and understand the terms of co-existence.

The civic dispositions of contemporary urban societies remain predominantly anthropocentric. They presuppose a categorical boundary between humans and other animals, normalising hierarchical relations and rendering nonhuman suffering peripheral to moral concern. These orientations are not sustained by abstract belief alone; they are reproduced through educational systems, economic structures, and urban environments that systematically confine animals to roles as commodities, companions, or background ecology. Transforming these dispositions is therefore integral to the project of the caring city. Legislation can prohibit cruelty and mandate ecological standards, but it cannot by itself cultivate the habits of perception through which citizens come to recognise nonhuman lives as morally present. That work happens in streets, parks, schools, and public spaces – which is why the design of urban environments is not merely an aesthetic or technical question but a political one.

The built environment contributes directly to this process. Architecture, urban design, and the organisation of public-space shape expectations about what forms of presence are legitimate and what are marginal. Consider a park designed exclusively for human recreation – with benches, artificial shade structures, water fountains for thirsty kids and adults, and perhaps ornamental birds confined in cages for public display. Compare it with a park intentionally structured for multispecies cohabitation, incorporating native vegetation, water features accessible to birds and insects, and shaded habitat for wildlife. The former treats nonhuman life as incidental or decorative; the latter incorporates it as a recognised dimension of shared space. These environments influence perception not through explicit instruction but by normalising certain forms of attention while rendering others peripheral. Benches that accommodate companion animals, water fountains accessible to people, birds, and other animals, and green roofs that support pollinators expand the range of recognised users of urban space. These designs communicate, not symbolically but materially, that nonhuman presence is anticipated within the city's design.

Such interventions accumulate into significant civic effects. When citizens encounter animals as ordinary co-inhabitants rather than as anomalies or nuisances, dispositions gradually shift. Designing transport systems that reduce wildlife collisions, incorporating animal ethics into public education, and linking food policy to questions of animal welfare together reconfigure what counts as civic responsibility. Care, in this sense, becomes a public competence.

The formative role of urban space is further reinforced through participatory governance. Citizen assemblies, community gardens, and local stewardship initiatives involve residents directly in the management of shared environments, cultivating habits of collective responsibility that extend beyond strictly human concerns. In Singapore, Community Gardens supported through the National Parks Board's Community in Bloom programme bring together volunteers across housing estates, schools, and organisations to cultivate edible plants, biodiversity-enhancing habitats, and ornamental species. Some gardens focus on food resilience and local self-provisioning; others prioritise habitat support for pollinators and native flora. Beyond their ecological functions, these gardens serve as hubs of collaboration and shared learning. Participants frequently report heightened awareness of ecological interdependence (Er, 2021; Tan & Neo, 2009). Through sustained engagement with soil, pollinators, and seasonal cycles, care becomes embodied knowledge: a practical orientation to shared vulnerability and dependence rather than an abstract ethical commitment.

The transformation of civic dispositions also requires unlearning. Anthropocentric socialisation has long trained citizens to disregard the presence of other species in public space. Stray animals, dead birds, or insects in domestic environments are treated as inconveniences to be eliminated rather than as indicators of shared ecological conditions. A caring city must actively recalibrate these reflexes, transforming habits of dismissal into practices of attentiveness. Integrating ecological literacy and animal ethics into school curricula, incorporating urban wildlife into public education, and developing civic rituals – memorials for nonhuman victims of environmental disasters, commemorations of extinct species – can frame vulnerability as a matter of collective responsibility rather than private sentiment (for example, see Wienhues & Donoso, 2024). As Thom van Dooren (2014) argues, practices of mourning can function as forms of ethical acknowledgement, situating loss within networks of interdependence and drawing attention to preventable harm.

Solidarity complements care by extending moral responsiveness into structural reform. Where care attends to immediate need and suffering, solidarity addresses the institutional arrangements that generate vulnerability in the first place. A citizen shaped by care responds to the plight of a stray animal; a citizen shaped by solidarity also seeks the regulatory and policy reforms that address abandonment, breeding practices, and municipal responsibility. The caring city requires both: sensitivity to concrete suffering and commitment to transforming the conditions that produce it – together, the interspecies civic ethos of a multispecies polity.

7 Conclusion

The caring city, as developed in this chapter, is best understood as a regulative ideal: a framework for orienting urban institutions toward interspecies justice. It does not promise harmony, but it redefines the political task of the city. In contrast to modern tendencies to materialise hierarchy, insulation, and exclusion, the caring city seeks to embed reciprocity and responsiveness in its spatial design, environmental metabolism, and civic practices. Justice becomes tangible not only in law but in waste systems, corridors, public spaces, and planning procedures.

This chapter has argued that urban injustice operates simultaneously through spatial design, environmental metabolism, and civic disposition – and that addressing it requires not isolated reforms but the transformation of all three dimensions together, constituting an interspecies urbanism in which self-development is structured as a shared condition of co-existence.

Such a transformation is unavoidably political. It confronts entrenched interests and dominant metrics of efficiency and growth. Yet it is also increasingly necessary. Climate instability, biodiversity loss, and the accelerating erosion of ecological thresholds reveal that the boundaries once imagined between the human city and the animal world are neither stable nor morally defensible. An urban order that externalises ecological harm ultimately undermines its own conditions of persistence.

Part II of the book has traced how care and solidarity can be institutionalised across the architecture of the polity: in constitutional design, in food systems, and in the city itself. The caring city brings these dimensions together. It translates normative commitment into material arrangement, metabolic responsibility, and civic ethos. In doing so, it demonstrates that interspecies justice is not an abstract aspiration but a question of how we build, regulate, and inhabit the shared spaces of political life.

Notes

1 Georges Franju's 1949 film *Le sang des bêtes*, which documented the routine work of Parisian abattoirs, was made at a moment when slaughterhouses were still relatively integrated into the urban fabric — a configuration that has since been progressively displaced to peripheral and rural sites (Franju, 1949)
2 For a comprehensive and careful account of the expansion of environmental justice beyond the human, see Wienhues (2020).
3 See Magaña (2022) for an excellent overview of the different ways in which animals can be politically represented.

References

Beatley, T. (2011). *Biophilic cities: Integrating nature into urban design and planning*. Island Press.

Bullard, R. D. (1990). *Dumping in Dixie: Race, class, and environmental quality*. Westview Press.

Clevenger, A. P., & Waltho, N. (2000). Factors influencing the effectiveness of wildlife underpasses in Banff National Park, Alberta, Canada. *Conservation Biology*, 14 (1), 47–56.

Clevenger, A. P., & Waltho, N. (2005). Performance indices to identify attributes of highway crossing structures facilitating movement of large mammals. *Biological Conservation*, 121 (3).

Cresswell, T. (1996) *In place/out of place: Geography, ideology, and transgression*. University of Minnesota Press.

Cresswell, T. (2004). *Place: A short introduction*. Blackwell.

Donaldson, S., & Kymlicka, W. (2011). *Zoopolis: A political theory of animal rights*. Oxford University Press.

Er, K. (2021). Transforming Singapore into a city in nature. *Urban Solutions*, 19, 68–77.

Fraser, N. (2008). *Scales of justice: Reimagining political space in a globalizing world*. Polity Press.

Franju, G. (Director). (1949). *Le sang des bêtes* [Film]. Forces et Voix de France.

Huber, D. et al. (2018). New highways and maintenance of large carnivore habitat continuity – a Croatian experience. *5th European Congress of Conservation Biology*. doi:10.17011/conference/eccb2018/107324.

Jordan, D. P. (1995). *Transforming Paris: The life and labors of Baron Haussmann*. Free Press.

Kennedy, C. et al. (2007). The changing metabolism of cities. *Journal of Industrial Ecology*, 11(2), 43–59.

Gellert, P., & Lynch, B. (2003). Mega-projects as displacements. *International Social Science Journal*, 55, 15–25. doi:10.1111/1468-2451.5501002.

Lefosse, D. et al. (2025). Biophilic Urbanism Across Scales: Enhancing Urban Nature Through Experience and Design. *Land*. doi:10.3390/land14051112.

Longcore, T., & Rich, C. (2004). Ecological light pollution. *Frontiers in Ecology and the Environment*, 2(4), 191–198.

Magaña, P. (2022). The political representation of nonhuman animals. *Social Theory and Practice*, 48(4), 665–690.

Masnavi, M. R. et al. (2019). Exploring urban resilience thinking for its application in urban planning: A review of literature. *International Journal of Environmental Science and Technology*, 16(1), 567–582. doi:10.1007/s13762-018-1860-2.

Schlosberg, D. (2007). *Defining environmental justice: Theories, movements, and nature*. Oxford University Press.

Shrader-Frechette, K. (2002). *Environmental justice: Creating equality, reclaiming democracy*. Oxford University Press.

Tan, L. H. H., & Neo, H. (2009). "Community in Bloom": local participation of community gardens in urban Singapore. *Local Environment*, 14(6), 529–539. doi:10.1080/13549830902904060.

Tronto, J. C. (1993). *Moral boundaries: A political argument for an ethic of care*. Routledge.

van Doren, B. et al. (2017). High-intensity urban light installation dramatically alters nocturnal bird migration. *Proceedings of the National Academy of Sciences*, 114, 11175–11180. doi:10.1073/pnas.1708574114.

van Doren, B. (2021). Drivers of fatal bird collisions in an urban center. *Proceedings of the National Academy of Sciences*, 118 (37), e2101666118 doi:10.1073/pnas.2101666118.

van Dooren, T. (2014). *Flight ways: Life and loss at the edge of extinction*. Columbia University Press.

Wenz, P. S. (1995). *Environmental justice*. State University of New York Press.

Wienhues, A. (2020). *Ecological justice and the extinction crisis: Giving living beings their due*. Routledge.

Wienhues, A. & Donoso, A. (2024). Reparations after species extinctions: An account of reparative interspecies justice. *Journal of Social Philosophy* 1, 21. doi:10.1111/josp.12584.

Wolch, J. (1998). Zoöpolis. In J. Wolch & J. Emel (Eds.), *Animal geographies: Place, politics, and identity in the nature-culture borderlands* (pp. 119–138). Verso.

Wolch, J. (2002). Anima urbis. *Progress in Human Geography*, 26, 721–742. doi:10.1191/0309132502ph400oa.

Yamagata, Y., & Sharify, H. (Eds.). (2018). *Urban resilience: A transformative approach*. Springer.

Young, I. M. (1990). *Justice and the politics of difference*. Princeton University Press.

Young, I. M. (2000). *Inclusion and democracy*. Oxford University Press.

Zellmer, A., & Goto, B. (2022). Urban wildlife corridors: Building bridges for wildlife and people. *Frontiers in Sustainable Cities*, 4, 954089. doi:10.3389/frsc.2022.954089.

PART III

Animal Justice Beyond Institutions

10

DIRECT ACTION AND THE LIMITS OF ANIMAL JUSTICE

1 Introduction

As argued in the second part of this book, the integration of care and solidarity in institutions lies at the core of a just multispecies community. A political order capable of recognising animals as subjects of vulnerability and justice requires more than moral recognition; it demands institutional arrangements that embed practices of protection, responsibility, and interdependence into the structure of collective life. Care and solidarity, in this view, are not merely ethical dispositions but constitutive political principles that shape how power, responsibility, and vulnerability are organised.

Yet the integration of these values into existing political systems is far from straightforward or assured. Modern societies, particularly those shaped by industrial capitalism, remain deeply resistant to the attitudes and values presupposed by an interspecies politics of care. Contemporary regimes of production, consumption, and growth rely upon a hierarchical separation between the human and the nonhuman, sustained by anthropocentric assumptions embedded in labour relations, legal conceptions of property, and prevailing expectations about the role of the state and the economy. Within such contexts, efforts to institutionalise care and solidarity confront structurally entrenched forms of anthropocentric privilege.

When these hierarchies persist, institutional transformation risks becoming insufficient or even complicit. Political institutions that systematically fail to protect or promote the self-development of those who inhabit them – human and nonhuman alike – cease to perform their justificatory function. Under conditions of oppression, the legitimacy of the state itself comes into question. This raises a pressing political problem: what forms of action remain available to pursue interspecies justice when the institutions tasked with realising it

DOI: 10.4324/9781003434856-13

instead reproduce its negation? Is there room for confrontational or disruptive practices that aim not to abandon political order altogether, but to restore its ethical foundations within a multispecies polity? This chapter takes up these questions by examining direct action as a form of restorative political practice that emerges precisely when institutional pathways are blocked.

The chapter proceeds in four steps. The next section develops a conceptual framework for understanding direct action and its normative status within a politics of care and solidarity. The following section introduces the Huntingdon Life Sciences campaign as a concrete site for analysing the ethical and political dimensions of direct action. The fourth section addresses the problem of violence in direct action and evaluates its compatibility with a restorative orientation grounded in care. The chapter concludes by clarifying the conditions under which disruptive political action may contribute to, rather than undermine, the long-term project of multispecies justice.

2 Direct Action

Direct action resists easy definition. It may nevertheless be understood, in broad terms, as a class of collective practices that intentionally circumvent, disrupt, or transgress formal legal structures, regulatory systems, or prevailing social norms in pursuit of social or political transformation. Historically, direct action has been deployed by a wide range of progressive movements not as a rejection of politics but as a means of contesting unjust political arrangements when institutional channels proved unresponsive or complicit. Sit-ins during the United States civil rights movement, the illegal occupation of workplaces by labour movements, the sabotage of infrastructure by anti-colonial struggles, and the blockade of environmentally destructive projects by climate activists all exemplify forms of direct action aimed at forcing moral and political questions back into public view when legality functioned as a shield for injustice.

The enduring tension between the performative character of direct action and the authority of the law reflects a deeper disjunction between legality and justice. Legal systems can operate as mechanisms of oppression as readily as they can serve as safeguards of liberty: they may criminalise legitimate dissent while normalising deeply unjust social arrangements. In such contexts, obedience to the law risks collapsing into complacency. When legal and democratic institutions function as mechanisms of inertia rather than transformation – when they systematically fail to protect the vulnerable or to challenge entrenched hierarchies – citizens may bear a political responsibility to act outside, and at times against, prevailing norms to realign political life with its ethical foundations.

It is within this broader tradition of political resistance that some animal activists have turned to direct action (see Tarrow, 2011). Faced with institutional frameworks that continue to permit, regulate, and normalise extensive harm to nonhuman animals, these actors interpret legality not as a neutral arbiter but as a structural expression of speciesist privilege. Direct action, in this context, is not conceived as a withdrawal from political responsibility but as an attempt to confront forms of injustice that remain insulated from critique precisely because they are legally sanctioned.

It is important, however, to distinguish between the roles played by animal activists and animal ethicists in advancing the cause of nonhuman animals. While direct action has long been embraced within activist circles as a necessary response to entrenched injustice, it remains far more controversial – and often explicitly rejected – within academic animal ethics. Much of the philosophical literature continues to prioritise rational persuasion as the primary, and sometimes exclusive, vehicle for political change.

A paradigmatic expression of this position can be found in the work of Peter Singer. In *Animal Liberation*, Singer emphasises the centrality of reasoned argument over confrontational tactics, insisting that the movement must win through the persuasive force of its moral case rather than through fear or coercion (Singer, 1990, pp. xii–xiii). This stance reflects a broader tendency within animal ethics not only to privilege normative argumentation and public reason over disruptive or extra-legal strategies, but to deny any political legitimacy to more confrontational forms of intervention.

At least in part, this resistance to direct action is grounded in a concern about democratic legitimacy. The underlying worry is that transgressing established, democratically affirmed norms can never be justified within a democratic polity. In this view, as long as institutional channels for participation, deliberation, and reform remain open, bypassing, disrupting, or violating laws, regulations, or social norms is necessarily illegitimate. Direct action thus appears not merely imprudent or counterproductive but democratically wrongful – an assault on the very procedures that secure political equality and collective self-rule.

This objection fails. It rests on the assumption that democratically established norms are, by virtue of their procedural origins, always legitimate – that is, aligned with the demands of justice. Yet this assumption is unwarranted. Political institutions may remain formally available while functioning de facto to neutralise just claims, to reproduce exclusion, or to insulate entrenched harms from meaningful contestation. Under such conditions, the mere availability of democratic procedures cannot perform the legitimating work the objection presupposes. Procedural openness is insufficient when institutional arrangements systematically block the realisation of justice.

Moreover, the appeal to collective resistance is not alien to democratic thought but internal to it. From Henry David Thoreau's refusal to fund slavery and war, to the suffragists' rejection of political exclusion, the civil rights movement's challenge to racial hierarchy, and environmental and decolonial struggles against extractivism, democratic progress has repeatedly depended on practices that transgressed existing legal norms. In each of these cases, disobedience expressed fidelity not to prevailing law but to a deeper conception of justice – one according to which moral obligation may, under certain conditions, supersede democratic legal authority.

As argued in Part I of this book, injustice does not arise solely, or even primarily, from isolated acts of individual wrongdoing, but from structural processes that systematically enable some forms of life while disabling others (Young, 2011). Individuals are implicated in these processes not necessarily because they directly harm others, but because they participate in and benefit from institutional arrangements that perpetuate oppression. This structural account of injustice correspondingly redefines political responsibility, extending it beyond compliance with existing norms to include a duty to transform the institutions that sustain unjust relations. Political responsibility encompasses not only obedience to law but, where necessary, principled resistance aimed at restoring the conditions under which justice – and self-development – can become possible.[1]

The legitimacy of the democratic state, I submit, does not depend primarily on consent, procedural legality, or formal inclusion alone, but on the state's capacity to secure the basic conditions necessary for a self-developing life. Political authority is justified, in this view, insofar as it sustains the social and institutional conditions under which individuals can develop, exercise, and revise their capacities over time. When this justificatory function is fulfilled, obedience to law carries moral weight; when it is not, the authority of law is correspondingly weakened.

Thus, when the state and the institutional structures it sustains persist in maintaining systems of oppression – understood here as systematic arrangements that block or undermine self-development – political legitimacy is forfeited. In such circumstances, compliance with existing legal norms can no longer be assumed to coincide with political responsibility. The persistence of formally democratic procedures is insufficient to restore legitimacy when substantive conditions for justice are absent.

This account of legitimacy admits a non-anthropocentric extension. If the state's foundational responsibility is to secure the basic conditions of self-development, and if nonhuman animals are among the beings whose lives can be organised in ways that enable or thwart such development, then the scope of legitimacy cannot be restricted to human subjects alone. When institutions systematically fail in this expanded foundational duty – by permitting, normalising, or facilitating the large-scale frustration of nonhuman self-development – citizens may be justified in acting outside the law or prevailing norms to restore minimal conditions of justice within a multispecies political community.

As shown throughout the previous chapters, this failure of institutional legitimacy is particularly evident in the case of nonhuman animals. Far from merely neglecting their interests, modern states actively institutionalise animals' oppression through legal and economic frameworks, including property regimes, agricultural subsidies, regulatory exemptions for research, and international trade policies. These arrangements do not represent accidental oversights but form part of a political and economic order that excludes animals from standing, representation, and effective advocacy.

Given the depth of this entrenchment, structural injustice toward animals cannot plausibly be remedied by persuasion alone or by incremental institutional transformation. Both law and political procedure are, in this domain, structurally designed to marginalise or silence the very beings whose interests are at stake. Under such conditions, the realisation of even the minimal requirements of justice – and, in particular, of the conditions necessary for nonhuman self-development – may depend on alternative forms of political action that do not rely exclusively on existing institutional channels. One such alternative is direct action.

The crucial point is that when democratic avenues are systematically foreclosed and oppression becomes normalised, disobedience need not signify rejection of political community. On the contrary, it can function as an act of civic fidelity rather than defiance – an expression of commitment to the ethical foundations of political life at precisely those moments when the state has departed from them and thereby forfeited its legitimacy.

The legitimacy of direct action is, however, conditional rather than general. Not every form of direct action is justified merely because institutional channels have proved inadequate. Legitimacy depends on the satisfaction of specific normative requirements, and I propose that direct action may be justified as an expression of interspecies political responsibility when three conditions are met: first, democratic channels for redress are systematically foreclosed; second, there is clear complicity in ongoing structural harm; and third, the action is guided by a genuine intent to restore relations of care rather than merely to disrupt or punish. Under these circumstances, disobedience is not merely permissible – it constitutes an enactment of political solidarity. Refusal to comply with entrenched norms, in such contexts, does not signal a breakdown of civic virtue but may instead express an obligation of justice grounded in fidelity to the ethical foundations of political life.

3 The Huntingdon Life Sciences Case

What follows develops this framework by examining how direct action can operate as an instrument of interspecies justice in practice. The campaign against Huntingdon Life Sciences (HLS) is a case in which conventional democratic channels proved persistently incapable of reforming entrenched systems of anthropocentric oppression. Rather than treating this campaign as an anomaly or excess, the discussion situates it within a framework of institutional failure and normative necessity, examining it through the lens of the three legitimacy conditions outlined above.

In December 2008, following a long campaign against HLS, a jury at Winchester Crown Court convicted four activists of conspiracy to blackmail.

The verdict marked the culmination of one of the most sustained and militant animal rights campaigns in the United Kingdom. Members of the Stop Huntingdon Animal Cruelty (SHAC) movement employed a broad repertoire of tactics – from sustained public protest and digital disruption to intimidation and acts of sabotage – aimed at dismantling the laboratory's financial and logistical support networks (Burns, 2008).

HLS, at the time Europe's largest contract research organisation conducting animal testing for the pharmaceutical and chemical industries, became the focus of a transnational mobilisation extending over a decade (1999–2014). SHAC adopted a decentralised strategy that targeted not only the laboratory itself but also the investors, insurers, suppliers, and corporate partners on which its operations depended. The objective was not merely symbolic or communicative protest but the imposition of economic and reputational costs sufficient to render participation in animal experimentation untenable.

The campaign's scale prompted an extraordinary state response. New legislative measures were introduced, policing practices expanded, and elements of animal rights activism were reframed within the language of extremism and security. Although many activists were prosecuted, imprisoned, or subjected to extensive surveillance, the campaign succeeded in isolating HLS financially and reputationally. Major investors, insurers, and clients withdrew, leading to changes in ownership, rebranding, and relocation.

The SHAC campaign demonstrates how direct action, when directed at the enabling conditions of oppression, can function as a form of political intervention rather than as a series of isolated acts against a single offending site. Its effectiveness depended on shifting pressure from the laboratory alone to the broader network of institutions that sustained its operations. From the standpoint of interspecies justice – which understands oppression as systemic rather than episodic – this strategic redirection is both analytically significant and normatively coherent.

This strategy makes visible that injustice in this domain cannot be reduced to the conduct of one firm. It is reproduced through interlocking legal, economic, and social arrangements that collectively sustain and normalise harm. Targeting HLS's financial and logistical partners can therefore be understood as an effort to redistribute accountability by imposing costs on those who enable, profit from, or shield oppressive practices. Action directed at supporting institutions is not ancillary but aims at interrupting the broader arrangements through which injustice is maintained.

The central issue is whether SHAC's strategies satisfy the aforementioned three legitimacy conditions: democratic foreclosure, complicity in ongoing structural harm, and restorative orientation.

The first condition – democratic foreclosure – is plausibly met. By the late 1990s, animal experimentation in the United Kingdom was not merely legally permitted but deeply embedded within an institutional nexus linking governmental policy, scientific authority, and commercial interest. Despite decades of lawful campaigning, parliamentary lobbying, and public protest, the core practices of vivisection remained largely insulated from effective democratic revision.[2]

A concrete illustration is provided by Barry Horne's hunger strikes in the late 1990s. Horne acted in response to the Labour government's abandonment of a 1997 manifesto commitment to establish a royal commission on animal experimentation (Whetton, 2026; The Independent, 1998). The pledge had generated expectations within the animal rights movement and was widely regarded as a viable democratic pathway for addressing concerns about vivisection. Once in office, the Blair government not only failed to honour the commitment but reinforced support for animal experimentation, effectively closing the channel of redress it had publicly invoked. Horne's hunger strikes should therefore be understood not as a premature bypassing of democratic procedure but as a response to its practical foreclosure. They illustrate how direct action may arise not in opposition to democracy itself but in response to the erosion of democratic accountability.

More broadly, by the time the SHAC campaign intensified, the abandonment of the promised royal commission was not simply a discrete policy reversal but indicative of a pattern in which procedural democracy continued to operate while substantive avenues for reform in this domain narrowed. Political legitimacy was sustained through formal mechanisms of representation and consultation, yet meaningful contestation of entrenched economic and scientific interests proved increasingly marginal. Regulatory oversight was limited, political incentives aligned with industry actors, and public concern was repeatedly deflected through appeals to scientific necessity. Taken together, these factors support the claim that institutional transformation was unlikely within any reasonable timeframe. Direct action, in this account, arose not from impatience with democratic delay but from a judgement that democratic accountability had become functionally inert in relation to the practices at issue.

Determining when foreclosure has occurred is not straightforward. In authoritarian regimes, it is clear; in democratic contexts, it is typically partial and domain-specific. The relevant question is not whether procedures formally persist but whether they retain meaningful transformative capacity in the area at issue. In the case of animal justice, there are strong reasons for the conclusion that they do not. The scale and durability of institutionalised harm toward nonhuman animals reflect a structural mismatch between the depth of the injustice and the limited transformative reach of existing legal and political mechanisms. The difficulty lies not merely in political delay but in the embeddedness of animal exploitation within economic, regulatory, and property regimes that resist fundamental reform.

The second condition – complicity in harm – is likewise plausibly satisfied. The injustice at issue was not produced by HLS in isolation but sustained through an interconnected network of financial, legal, scientific, and commercial actors whose coordinated participation rendered animal experimentation economically viable and socially entrenched. Banks provided credit and financial services; insurers underwrote operational risks; suppliers furnished equipment and animals; universities and research institutions outsourced ethically contentious experimentation while preserving their public reputations (Treanor et al., 2001). Taken together, these actors constituted a distributed structure of facilitation in which responsibility was diffused across institutional roles, thereby reducing the visibility and perceived accountability of any single participant.

At least at the level of declared strategy, SHAC aimed to extend pressure outward along this chain of facilitation, seeking to render background participation in animal experimentation both visible and materially costly. Where oppression is sustained through dispersed yet coordinated forms of participation, legitimacy cannot be evaluated solely by reference to the immediacy of harm at a single site. It must also consider whether direct action meaningfully engages the wider configuration of complicity through which that harm is reproduced. To the extent that SHAC's strategy tracked this architecture of responsibility rather than treating HLS as an isolated offender, the second condition is substantially met.

The third condition – restorative orientation – introduces a more demanding constraint. Even where direct action tracks networks of facilitation rather than isolated actors, its legitimacy depends on whether it remains directed toward the reconstruction of just relations or whether it foreseeably devolves into dynamics that undermine that objective.

In the SHAC campaign, strategies designed to expose and disrupt complicity sometimes involved escalating pressure on secondary and tertiary actors through fear, intimidation, and social exclusion. This raises a question that cannot be fully resolved here: whether such tactics can remain compatible with a restorative orientation, or whether they foreseeably corrode the relational conditions that restoration requires. From a care-ethical standpoint, confrontation may be necessary to interrupt entrenched harm – but whether violent or coercive forms of confrontation can satisfy the restorative condition is a question the following section addresses directly.

The evaluative task, therefore, is not simply to catalogue tactics but to assess whether SHAC's practices sustained a forward-looking commitment to institutional repair or, in some instances, reproduced patterns of oppression and moral hardening – the erosion of the empathetic and relational capacities that a restorative orientation requires.

4 Direct Action and Violence

If the restorative condition constrains the aims and moral orientation of legitimate direct action, it also raises a harder question: can direct action involving violence ever satisfy it? The issue is not whether violence is effective but whether violence can be reconciled with an orientation toward care and solidarity. Strategies that seek to render complicity materially costly often involve escalating forms of coercive pressure, thereby destabilising the boundary between forceful denunciation and violence and making that boundary normatively salient. In this section I examine whether violent forms of direct action can remain compatible with a care-ethical commitment to rebuilding just relations, or whether violence – given its characteristic tendency to generate fear, resentment, and retaliatory escalation – tends to corrode the moral foundations the restorative condition seeks to protect. The question is particularly acute in the case of SHAC, whose tactics were widely characterised as violent, and where it would be disingenuous to deny that some crossed conventional thresholds of violence.

The argument developed here finds important support in Candice Delmas's philosophical account of principled resistance (Delmas, 2018). She challenges the idealisation of civil disobedience as necessarily non-violent, public, and decorous, arguing that principled resistance to injustice may legitimately take forms that conventional political theory regards as impermissible – what she terms "uncivil" disobedience. Crucially, Delmas grounds this not in a permissive claim, for example, that such actions are merely tolerable, but in an obligatory one: that citizens have, under specifiable conditions, a moral duty to resist injustice, a duty she derives from multiple mainstream theories of political obligation, including the natural duty of justice, fair play, and Samaritan duties. The present argument extends this framework to the interspecies domain. If the duty to resist injustice is grounded in principles that operate independently of the identity of the oppressed – as Delmas's ecumenical approach implies – then there is no principled basis for confining that duty to cases of human oppression. Where animals are subjected to structural injustice that institutional channels have failed to address, the same duty applies.

This extension is not merely theoretical. The history of animal advocacy contains movements and individuals who arrived, through their own moral reasoning, at precisely this conclusion – that the scale and intractability of animal oppression generated not merely a permission but an obligation to act outside and against prevailing norms. The moral logic underpinning SHAC's actions was already visible in earlier British animal activism, including the milieu in which Barry Horne operated (Ellefsen, 2018). Activists associated with the Animal Liberation Front often understood themselves as responding to an ongoing moral emergency: a social order structured around systematic animal killing that appeared to suspend ordinary constraints on political action. Within this framework, moral outrage did not merely justify protest; it generated a sense of obligation to intervene.[3]

Yet the obligation Delmas identifies does not settle the question of whether violent action satisfies the restorative condition.[4] Even where violence is morally required in principle, its practical effects may undermine the very goals that obligation is meant to serve. This concern is developed empirically by N.

Müller and Spang (2024), who argue that violent activism is counterproductive in most cases: rather than weakening the regimes that sustain animal oppression, it tends to trigger self-reinforcing reactions that entrench those regimes further. When direct action is framed as a response to moral horrors of such magnitude that conventional limits lose authority, violence risks being recast not as a tragic remainder but as an absolute demand – eroding the critical distance within which the restorative condition can function as a genuine normative limit.

It is precisely this combination of normative erosion and practical failure that makes it necessary to ask whether violence can remain compatible with an orientation toward repair. Violence motivated by genuine moral concern for animals may undermine the very goals it sets out to serve – unless it remains constrained by a framework that distinguishes between intervention oriented toward halting ongoing harm and violence that devolves into oppression or retribution.

In the context of environmental movements – and, by extension, wherever institutions fail to protect the vulnerable from ongoing structural harm – political theorist Ross Mittiga (2024) has argued that dominant theories of civil disobedience tend to construe disobedience primarily as a communicative practice aimed at persuasion, thereby overlooking what he terms harm-preventative civil disobedience: action directed at obstructing practices that pose grave and ongoing threats, particularly where governments fail in their duty to protect collective safety. In such circumstances, disobedience functions not as a symbolic protest but as a form of intervention intended to halt harm already underway.

From a care-ethics perspective, Mittiga's reframing is compelling because it foregrounds responsiveness to vulnerability and the urgency of acting in the face of imminent harm. Yet it also exposes a tension internal to care itself. The commitment to protect those at risk must remain compatible with preserving the relational conditions – trust, reciprocity, accountability – through which repair becomes possible. Preventative intervention cannot simply override the normative constraints that make restoration meaningful. Legitimacy cannot therefore be assessed solely by reference to outcomes; it must also be evaluated in terms of how tactics reshape present and future relationships among agents, including victims, perpetrators, and facilitators of harm.

Practices grounded primarily in fear, humiliation, or oppression risk eroding the very capacities for mutual recognition and responsibility that restorative politics seeks to cultivate. This tension is particularly acute in campaigns such as SHAC, whose preventative orientation toward stopping harm to animals situates them squarely within Mittiga's framework while raising a question his account does not address: can a politics grounded in care legitimate violence without undermining the relational foundations that justify preventative intervention in the first place?

Seen in this light, the SHAC campaign appears not primarily as an attempt to persuade public opinion but as a deliberate effort to interrupt an ongoing structure of oppression. Its attempts to bankrupt the clients, insurers, and financial partners of HLS can be understood, adapting Mittiga's language, as an oppression-preventative strategy: targeting the economic and logistical architecture through which harm to nonhuman animals is systematically reproduced. This framing holds insofar as these violent actions are understood as being directed toward agents of facilitation – banks, laboratories, logistics firms, suppliers – in a context of structural injustice in Iris Marion Young's sense: a web of interdependencies through which oppression persists without any single actor exercising decisive control. Framed in this way, SHAC's strategy aligns with a preventative logic aimed at disabling entrenched harm.

A second question remains: whether the particular tactics employed are compatible with the restorative orientation. SHAC's acts can be understood as attempts to interrupt ongoing arrangements that reproduce animal vulnerability and systematically exclude nonhuman animals from political consideration – and, extending Mittiga's claim that citizens may become final guarantors of safety when institutions fail (Mittiga, 2024, p. 78), such actions may at times require violence. Yet their legitimacy does not turn on the mere presence or absence of violence. The decisive question is whether the means employed remain compatible with preserving the conditions under which repair of institutions, relations, and the political community itself can become possible. Where tactics foreseeably undermine those relational conditions, their oppression-preventative logic cannot by itself secure legitimacy.

Any account of legitimate escalation must remain context-sensitive. The relevant threshold cannot be fixed in abstraction from concrete circumstances. In other words, whether oppression-preventative action crosses into illegitimate violence depends on a set of contextual factors: the severity and imminence of the harm resisted, the exhaustion of non-violent alternatives, the proportionality of the response, and the foreseeable effects of action on the moral fabric of the political community. From the standpoint of restoration, the decisive question is not simply whether violence is employed but whether its use remains compatible with preserving the possibility of repairing relationships rather than entrenching injustice. Violence that indiscriminately spreads terror, humiliates opponents, or foreseeably inflicts severe bodily harm corrodes the conditions under which a shared political future can be rebuilt and therefore fails the restorative test.

By contrast, violence directed at material infrastructures – damaging facilities or imposing economic constraints on institutions that sustain oppressive practices – does not automatically violate the restorative condition, provided it aims to interrupt ongoing structural injustice rather than merely to express condemnation or inflict fear. Where measures are proportionate, carefully targeted, and responsive to changing circumstances, they may remain compatible with a forward-looking commitment to repair.

4.1 Non-Violent Incivilities

Non-violent incivilities constitute a category of action that falls below the threshold of violence but above the threshold of conventional civil conduct. Joel Feinberg's concept of harmless wrongs is useful here (Feinberg, 1988). Harmless wrongs are actions that may offend, disturb, or symbolically violate norms without directly inflicting injury. In contexts of entrenched structural injustice and institutional foreclosure, certain forms of disruptive expression – offensive language, confrontational performance, symbolic intrusion – may fall within the limits of legitimate direct action when they contribute, even indirectly, to interrupting oppressive arrangements and remain consistent with restorative aims. Though primarily expressive, such acts may unsettle complacency, expose complicity, and catalyse withdrawal from harmful practices. The SHAC campaign employed tactics of this kind alongside more forceful measures, and their legitimacy should be assessed by the same criteria: not whether they transgress conventional standards of civility, but whether they meaningfully disrupt ongoing structures of oppression without foreclosing the relational conditions required for solidaristic repair.

Understood in this way, the restorative condition does not impose an absolute prohibition on all uses of force or uncivility. It functions instead as a normative filter operating in conjunction with the oppression-preventative requirement: escalation must both interrupt structural harm and preserve the possibility of rebuilding a political community capable of care and solidarity across species boundaries. Direct action that foreseeably destroys the conditions for future coexistence fails this test, regardless of immediate success. Direct action that disrupts entrenched injustice while preserving the relational space for repair, whether through non-violent incivility or constrained recourse to force, can be interpreted as an attempt to restore political legitimacy by reasserting moral commitments the legal order has failed to honour.

5 Conclusion

This chapter has argued that direct action can, under specific conditions, function as a legitimate mode of interspecies political responsibility. This conclusion does not dismiss the value of democratic procedure. On the contrary, direct action is treated as exceptional precisely because democratic institutions ordinarily provide the most stable framework for coordinating collective life. The claim is not that legality is irrelevant to legitimacy, but that legality can forfeit its normative authority when it becomes structurally detached from the protection of vulnerable beings whose lives it governs.

The three conditions presented in this chapter – democratic foreclosure, structural complicity, and restorative orientation – do not license disobedience indiscriminately. They operate as constraints that sharply narrow the circumstances under which the circumvention of law may be justified. Together they articulate a demanding standard: direct action becomes legitimate only where institutional mechanisms have ceased to offer a plausible path to justice, where oppression is sustained through distributed structures that evade ordinary accountability, and where action remains oriented toward the reconstruction of caring and solidaristic relations rather than their erosion. Understood in this way, direct action is neither a rejection of politics nor a withdrawal from shared norms. It is a claim about what fidelity to those norms requires when their institutional embodiment collapses.

The case of Huntingdon Life Sciences illustrates both the force and the fragility of this position. It demonstrates how structural injustice can generate forms of activism intelligible as attempts to restore political legitimacy rather than merely to express outrage. At the same time, it shows how easily preventative intervention can escalate into tactics that risk reproducing the very dynamics of oppression that direct action seeks to oppose. The restorative condition thus functions as the moral hinge on which legitimacy turns: it insists that even justified confrontation remains answerable to the long-term task of rebuilding relations of recognition and responsibility across species lines.

The implications extend beyond this case. If animal justice requires institutions capable of responding to vulnerability wherever it appears, then the persistence of direct action movements should be read not merely as a law-enforcement problem but as a diagnostic indicator of normative deficit. Where citizens repeatedly resort to extra-legal intervention to prevent ongoing harm, the central question is not only whether they have violated the law but whether the law has ceased to perform the protective function that grounds its authority. In this sense, direct action occupies an ambivalent yet revealing position within a multispecies polity, marking the point at which care confronts the limits of institutional design.

A just political order cannot assume that confrontation can be abolished. Confrontation emerges wherever institutions fail to command moral authority, and the appropriate response is not repression but the transformation of the conditions that make it necessary. Yet contemporary institutional arrangements continue to expose vulnerable beings to systematic harm, eroding the legitimacy they presuppose. When authority becomes detached from the protection of those it governs, confrontation – even if violent – ceases to be an aberration and becomes a predictable feature of political life. Under such conditions, the possibility of legitimate direct action is not an accident at the margins of politics but a sign that political institutions remain ethically incomplete.

Notes

1 This claim can be understood as a development of what John Rawls describes as a natural duty to support just institutions – extended here to include the duty to oppose and reform institutions that systematically fail to meet that standard. Rawls

himself developed an account of civil disobedience that grounds resistance in the same natural duty (Rawls, 1971, §55–59), though within bounds that the present chapter argues are insufficient for cases of structural anthropocentric injustice.

2 For a helpful overview of the development of animal experimentation in the UK and contemporary statistics on animal use in scientific research, see Animal Free Research UK (n.d.).

3 Contemporary media accounts frequently described ALF activists as understanding animal exploitation in catastrophic moral terms. The obituary of Horne published in *The Guardian* characterises this outlook as a form of absolutist moral urgency – the rhetoric is journalistic, but historically revealing.

4 It must be noted that the very concept of violence is contested (Bernstein et al., 2018). As used here, violence refers not only to direct bodily harm but to the exertion of force that violates persons or the domains within which their agency is normatively secured – including property destruction and coercive intrusion (such as repeated home visits or surveillance). Steven Best adopts a narrower account, restricting violence to acts such as assault, kidnapping, torture, and assassination (Best, 2014, p. 60), and argues that broader definitions are frequently deployed to delegitimise dissent. He further rejects accounts that include property damage, arguing that they blur morally significant distinctions between harm to living beings and damage to objects (Best, 2014, p. 62). There is force in both observations – the language of violence is often strategically mobilised, and the distinction between harm to persons and damage to property is morally significant. The present argument does not deny this distinction but questions whether it exhausts the concept of violence. Whether an act constitutes violence is a descriptive question about its form and foreseeable effects, not a verdict on its moral status – a distinction the main text presupposes throughout.

References

Animal Free Research UK. (n.d.). UK law. www.animalfreeresearchuk.org/uk-law/.

Bernstein, J. M. et al. (Eds.). (2018). *Political concepts: A critical lexicon*. Fordham University Press.

Best, S. (2014). *The politics of total liberation: Revolution for the 21st century*. Palgrave Macmillan.

Burns, J. (2008, December 24). Britain: Animal activist convicted. *The New York Times*. www.nytimes.com/2008/12/24/world/europe/24briefs-ANIMALACTIVI_BRF.html.

Delmas, C. (2018). *A duty to resist: When disobedience should be uncivil*. Oxford University Press.

Ellefsen, R. (2018). Performing and policing transgressive protest: A relational approach to the SHAC-HLS conflict in Britain (1999–2014) [Doctoral thesis, University of Oslo]. www.researchgate.net/publication/338792220.

Feinberg, J. (1988). Harmless wrongdoing (Vol. 4 of The moral limits of the criminal law). Oxford University Press.

Mittiga, R. (2024). *Climate change as political catastrophe: Before collapse*. Oxford University Press.

Müller, N., & Spang, F. (2024). The counterproductiveness argument against animal rights violence. *Journal of Applied Philosophy*, 41 (5), 827–845.

Singer, P. (1990). *Animal liberation* (2nd ed.). New York Review of Books.

Tarrow, S. (2011). *Power in movement: Social movements and contentious politics* (3rd ed.). Cambridge University Press.

The Independent. (1998, November 6). Animal welfare is a good cause but it needs no martyrs. *The Independent*. www.independent.co.uk/arts-entertainment/leading-article-animal-welfare-is-a-good-cause-but-it-needs-no-martyrs-1189780.html.

Treanor, J. et al. (2001, January 18). Huntingdon Life: Facing collapse in 36 hours. *The Guardian*. www.theguardian.com/business/2001/jan/18/uknews1.

Whetton, N. (2026, February 9). Animal rights protesters targeted by a Labour government … again. *LSE British Politics and Policy*. https://blogs.lse.ac.uk/politicsandpolicy/animal-rights-protesters-targeted-by-a-labour-governmentagain/.

Young, I. M. (2011). *Responsibility for justice*. Oxford University Press.

11

THAT WAS MY PIG

This book began with two real stories. The first concerned Carolina, a 15-year-old girl from Guatemala, packing cereal boxes in the middle of the night in an American factory, thousands of miles from home, alone. The second concerned a group of rhesus macaques, taken from a monastery in northern Thailand, transported through falsified supply chains, and deposited in American biomedical laboratories to live and die in sterile cages. These stories were offered not as illustrations of individual cruelty but as windows onto a structural condition: the organisation of human and nonhuman lives within systems that convert bodies into instruments, economic value, and resources to be managed and exploited. The book has argued that what Carolina and the macaques share is not merely suffering but oppression, a specific and serious form of injustice sustained not by the malice of particular individuals but by the institutional arrangements, cultural norms, and economic logics that organise their lives from the outside, independently of anyone's intentions.

There is a third story, less visible than the first two, that the book has not yet told directly but toward which its argument has been moving throughout. It comes from a testimony given by a United States military corpsman describing an advanced trauma treatment programme he had taken before deploying (Chivers, 2006). The instructors gave him an anaesthetised pig. His task was to keep the animal alive. Then the test began.

> "The idea is to work with live tissue," he said. "You get a pig and you keep it alive. And every time I did something to help him, they would wound him again. So you see what shock does, and what happens when more wounds are received by a wounded creature."
>
> "My pig?" he said. "They shot him twice in the face with a 9-millimetre pistol, and then six times with an AK-47 and then twice with a 12-gauge shotgun. And then he was set on fire."

DOI: 10.4324/9781003434856-14

> "I kept him alive for 15 hours," he said. "That was my pig."
>
> *(Chivers, 2006)*

It is difficult to read this testimony without recognising in it something that exceeds the horror of any individual act. What is being described is not cruelty in the ordinary sense – not the sadism of a particular individual but the operation of a structure. The military institution has designed a system in which the pig's suffering is the medium through which the corpsman's competence is produced. The corpsman's role is to repair what the instructors destroy, and then to watch it destroyed again. He is not positioned as a perpetrator, but as a carer – and yet the caring is inseparable from the wounding, because the wounding is what the caring is for. The structure converts his capacity for attentiveness and responsiveness, precisely the capacities this book has associated with care as a political value, into instruments of an institution organised around the production of military utility.

What this image reveals, with an almost unbearable clarity, is the profound entanglement of human and nonhuman oppression. The pig and the corpsman are both captured within the same structure, which uses the suffering of one to shape the other. The pig is subjected, simultaneously, to exploitation, powerlessness, marginalisation, cultural imperialism, and violence – all five faces of oppression – in a setting that is legally sanctioned, institutionally organised, and culturally invisible. The corpsman, meanwhile, undergoes a different but related form of structural harm: his moral and psychological capacities are distorted by an institution that requires him to suppress his natural response to suffering, to measure his own competence through the medium of another being's pain, and to internalise a logic of instrumental use that implicates his humanity in the pig's destruction. He is not the pig's oppressor in any straightforward sense. He is, in a different register, a fellow captive of the structure that organises them both.

This is the book's deepest claim. The oppression of animals and the oppression of humans are not parallel injustices running on separate tracks. They are structurally entangled: produced by the same institutional logics, sustained by the same cultural norms, and reproduced through the same political and economic arrangements. To understand one fully is to understand the other more clearly. To resist one seriously is to confront the structures that make both possible.

The first part of the book traced this entanglement across three distinct but interconnected spheres of structural injustice. Each sphere illuminated a different face of the same underlying logic, and together they establish something that no single chapter could establish alone: that the oppression of animals is not a localised or exceptional phenomenon but a pervasive and constitutive feature of the political and economic order in which we live.

Chapter 2 established the theoretical foundation. Drawing on Iris Marion Young's account of structural oppression, it showed that animals in industrialised societies are subjected to all five faces of oppression – exploitation, marginalisation, powerlessness, cultural imperialism, and violence – through institutional arrangements that operate independently of individual intent and many times with the full authority of law. The central harm is the obstruction of self-development: the systematic denial, through structural position, of the conditions under which beings can develop and exercise their capacities. This is not a harm that can be addressed by reducing individual cruelty. It requires transforming the structures that produce it.

Chapter 3 deepened this analysis by showing how ableism and speciesism share a common logic of valuation. Capitalist societies measure the worth of bodies – human and nonhuman alike – according to norms of productivity, efficiency, and economic contribution. Bodies that fail to meet these norms are rendered disposable: culled, discarded, or simply allowed to die. The poultry industry provided the clearest illustration of this logic at work, but the chapter's argument extended beyond it – to the broader political economy that organises vulnerability as a resource and treats the elimination of the non-productive as a technical solution rather than a moral catastrophe.

Chapter 4 traced this argument across the site of migration, showing how the militarised border simultaneously endangers human migrants and fragments the ecosystems on which nonhuman animals depend – binding their vulnerability within the same regime of containment and exclusion. But the border is not the endpoint of this process; it is its point of production. The legally precarious, socially invisible human subjects generated at the frontier are distributed into the labour markets of industrial capitalism – including, paradigmatically, the slaughterhouses and dairy farms where migrant workers and animals are jointly governed through precarity, violence, and disposability. The border and the slaughterhouse are not separate domains but consecutive moments within the same political economy: one that generates disposable human labour at the frontier and deploys it in the service of animal killing at the interior.

Chapter 5 extended this analysis to the global scale, examining how the expansion of industrial palm oil production displaces the Batek people of Malaysia and destroys the habitat of the Borneo elephants who share their world. Here, the entanglement of human and nonhuman oppression is laid bare with particular clarity: the same global industry, driven by the same economic logics of extraction and profit, renders Indigenous communities and nonhuman animals simultaneously disposable – erasing their forms of life, severing their relationships to land and to each other, and converting their shared world into a commodity. What a politics attentive only to human suffering cannot see is that these are not two separate injustices requiring two separate responses. They are a single structural process whose full dimensions only become visible when the species boundary ceases to function as a limit of moral and political concern.

What these four chapters establish, taken together, is that the oppression of animals is not an anomaly within an otherwise just order but a structural feature of the order itself, one of the mechanisms through which capitalism, border regimes, and ableist norms of productivity reproduce themselves, justify their hierarchies, and naturalise their violence.

Against this logic, the second part of the book developed a normative response organised around care and solidarity, not as sentimental additions to a theory of justice but as its structural conditions in a world shared across species lines. To care, in the sense developed here, is not to offer compassion to victims of misfortune but to take institutional responsibility for dismantling the arrangements that make vulnerability political. To act in solidarity is not to speak for animals but to resist with them, challenging, through institutions and practices, the hierarchies that define their lives as expendable.

Care and solidarity are not abstract values; they are principles with concrete institutional implications, and those implications are demanding. A caring constitution recognises all beings capable of self-development as subjects of political governance rather than objects of regulatory discretion – not by attributing to them capacities they lack but by ensuring that their interests, vulnerabilities, and capacities are structurally incorporated into the processes through which collective decisions are made. A caring food system treats animal exploitation not as an unfortunate but inevitable cost of feeding human populations but as a political choice, one already being contested and partially superseded by alternative models of production that decouple nourishment from oppression. A caring city recognises nonhuman life as a constituent dimension of urban existence, redesigning its spaces, metabolic systems, and civic culture around the demands of multispecies coexistence rather than the convenience of a single species.

What these institutional visions share is a common premise: that justice is not a constraint to be applied after the fact to arrangements organised around other priorities, but a threshold condition for the legitimacy of those arrangements in the first place. Systems that require oppression to function cannot be rendered just through incremental reform. They must be replaced by systems whose design reflects, from the outset, the moral standing of all those they affect. This is a demanding standard, but it is the standard that follows inescapably from taking seriously the claim that animals are subjects of justice.

The third part of the book confronted the possibility that this institutional transformation may be blocked by the deliberate resistance of those who benefit from existing arrangements. Democratic institutions can function as mechanisms of entrenchment as readily as mechanisms of reform, and when they do, the question of what remains available to those committed to justice becomes urgent and unavoidable. The book has answered that direct action – including, under specific and demanding conditions, the use of force against material infrastructures of oppression – may be legitimate when institutional channels have been foreclosed, when structural complicity is pervasive, and when action remains oriented toward the restoration of the conditions under which a just political order becomes possible. This is not a licence for indiscriminate resistance. It is a recognition that the authority of law depends on its capacity to protect the vulnerable, and that when law forfeits that capacity, it forfeits the deference it is owed.

It would be a failure of intellectual honesty to conclude without acknowledging what this book has not resolved, and what, given its scope and ambition, it could not have resolved within a single volume.

The most pressing unresolved question concerns genuine conflict. This book has argued throughout that the oppression of humans and the oppression of animals are structurally entangled – that attending to one requires attending to the other, and that the political imagination capable of addressing both is more expansive than either alone. But entanglement is not identity, and the interests of humans and animals do not always converge. There are cases – genuine, difficult, and not easily dismissed – in which the needs of vulnerable human communities and the needs of animals come into real tension: subsistence practices that depend on animal use, medical research that cannot yet be replaced by alternatives, triage decisions in disaster contexts where the allocation of scarce resources forces choices between human and nonhuman lives that no general principle can resolve in advance. This book has touched on some of these and other tensions but has not developed a systematic framework for adjudicating them. That framework remains a task for future work, one that will require the kind of careful, contextually sensitive analysis that a book organised around structural critique can only partially provide.

A second limitation concerns democratic legitimacy. The multispecies polity this book envisions is one in which the claims of nonhuman animals are incorporated into political institutions and constitutional frameworks. But the process of incorporation is itself a political act, performed by humans, on behalf of beings who cannot consent to it or contest it in conventional ways. This raises a genuine worry: that even the most well-intentioned project of animal justice risks reproducing, in a different register, the paternalism it sets out to challenge, substituting human judgements about animal interests for the structural subordination those judgements are meant to contest. This book has no fully satisfying answer to this worry. What it can say is that the absence of a perfect solution is not a reason to preserve an unjust status quo, and that the development of institutional mechanisms – animal advocates, ecological ombudspersons, structured forms of proxy representation – that bring animal perspectives as close as possible to the centre of political deliberation is a more honest response to the problem than continued exclusion. But the tension between representation and paternalism is real, and any serious theory of interspecies justice must continue to grapple with it.

These limitations do not undermine the book's central argument. They locate it within a larger and ongoing inquiry – one that this book hopes to have advanced, without pretending to have completed.

The interspecies polity this book envisions is not a utopian ideal of harmony between species that have always been in complex and sometimes violent relation. It is a framework of accountability, a political order in which the capacity to suffer, to develop, and to be harmed by structural arrangements generates claims that institutions are obligated to recognise and respond to, regardless of species. Such a polity does not stand apart from nature, imagining itself as a purely human construction imposed upon a passive world. It is situated within it, constituted by the ecological relationships, dependencies, and vulnerabilities that bind human and nonhuman lives together in ways that no political boundary can dissolve.

The construction of such a polity is already underway: in sanctuary farming and agroecological cooperation, in legal campaigns for animal personhood, in constitutional provisions recognising animal rights and rights of nature, in direct action movements that refuse to wait for institutional permission to resist ongoing harm, and in the theoretical and practical work of the scholars, activists, and communities whose efforts this book has tried to build upon and extend.

What remains is the hardest part: not the articulation of principles but their institutionalisation; not the recognition of injustice but the transformation of the arrangements that produce it. That transformation will be resisted by those who benefit from existing hierarchies, by institutions designed to reproduce them, and by the cultural norms that render oppression invisible by making it ordinary. It will require patience, solidarity, and a willingness to follow the argument wherever it leads, even when the destination is uncomfortable and the path is contested.

Carolina is still out there, somewhere, building a life within the constraints that structural injustice has imposed upon it. The macaques are still in their laboratories, or they are not – their lives are short, and the system that produced them continues undisturbed. The pig died not long after the fifteen hours ended. The corpsman lives with what he was trained to do and what that training required him to become. These are not stories with resolutions. They are invitations to see the structures that connect them, to understand what justice would require in response, and to begin, however partially and imperfectly, the work of building something different.

References

Chivers, C. J. (2006, November 2). Tending a fallen Marine, with skill, prayer and fury. *The New York Times*. www.nytimes.com/2006/11/02/world/middleeast/tending-a-fallen-marine-with-skill-prayer-and-fury.html.

INDEX

For Product Safety Concerns and Information please contact our
EU representative GPSR@taylorandfrancis.com Taylor & Francis
Verlag GmbH, Kaufingerstraße 24, 80331 München, Germany